Cities, Climate Change, and Public Health

Cities, Climate Change, and Public Health

Building Human Resilience to Climate Change at the Local Level

Ella Jisun Kim

ANTHEM PRESS

Anthem Press
An imprint of Wimbledon Publishing Company
www.anthempress.com

This edition first published in UK and USA 2020
by ANTHEM PRESS
75–76 Blackfriars Road, London SE1 8HA, UK
or PO Box 9779, London SW19 7ZG, UK
and
244 Madison Ave #116, New York, NY 10016, USA

British Library Cataloguing-in-Publication Data
A catalogue record for this book is available from the British Library.

Library of Congress Cataloging-in-Publication Data
Library of Congress Control Number: 2019955639

ISBN-13: 978-1-78527-324-7 (Hbk)
ISBN-10: 1-78527-324-8 (Hbk)

This title is also available as an e-book.

To Miro

CONTENTS

List of Figures and Tables		ix
Acknowledgments		xi
Chapter 1	Introduction	1
Chapter 2	Cities, Climate Change, and Public Health	5
Chapter 3	Collective Climate Adaptation at the Local Level	13
Chapter 4	Framing Climate Change as a Public Health Issue	21
Chapter 5	Role-Play Simulations	31
Chapter 6	Digital Games	49
Chapter 7	Role-Play Simulations versus Digital Games	65
Chapter 8	Discussion and Policy Recommendations	73
Appendix A.	*Issue Framing Vignettes*	83
Appendix B.	*Issue Framing Survey Questions*	85
Appendix C.	*Role-Play Simulation Before-Survey Questions*	89
Appendix D.	*Role-Play Simulation After-Survey Questions*	93
Appendix E.	*Role-Play Simulation Facilitator Questionnaire*	97
Appendix F.	*Role-Play Simulation Debriefing Script*	99
Appendix G.	*Role-Play Simulation Post-Game Interview Questions*	101
Appendix H.	*Role-Play Simulation Workshop Sample Characteristics*	103
Appendix I.	*Digital Game Before-Survey Questions*	105
Appendix J.	*Digital Game After-Survey Questions*	107
Appendix K.	*Digital Game Post-Game Interview Questions*	111
Appendix L.	*Digital Game Sample Characteristics*	113
References		115
Index		129

FIGURES AND TABLES

Figures

5.1	Overview of Face-to-Face Role-Play Simulation and Data Collection	36
5.2	Effects of RPS on Different Types of Learning by Political Viewpoint	44
6.1	Home Screen of Digital Game	51
6.2	Phase 1 of Martin	52
6.3	Phase 2 of Martin	53
6.4	Phase 3 of Martin	53
6.5	Effects of Digital Game on Different Types of Learning by Political Viewpoint	58
6.6	Proactive Learning: Checking the Heat Index	60
6.7	Proactive Learning: Checking the Air Quality Index	60

Tables

2.1	Human Health Effects of Climate Change	6
2.2	The Five Steps of BRACE: Building Resilience Against Climate Effects	9
2.3	Checklist for Integrating Health into Climate Planning	10
4.1	Main Effects of Framing on Key Dependent Variables	26
5.1	Main Effects of RPS on Key Dependent Variables	41
5.2	Main Effects of RPS on Key Dependent Variables by Political Viewpoint	42
5.3	Effects of RPS on Different Types of Learning by Political Viewpoint	43
5.4	Prioritization of Local Climate Risks in Climate Adaptation, Before and After RPS	45
5.5	Prioritization of Policy Approaches to Manage Climate Risks, Before and After RPS	45
6.1	Main Effects of Digital Game on Key Dependent Variables	56

6.2 Main Effects of Digital Game on Key Dependent Variables by Political Viewpoint 57

6.3 Effects of Digital Game on Different Types of Learning by Political Viewpoint 58

6.4 Prioritization of Local Climate Risks in Climate Adaptation, Before and After Digital Game 59

7.1 Role-Play Simulation versus Digital Game Demographics 66

7.2 Main Effects of RPS versus Digital Game on Key Dependent Variables 67

7.3 Main Effects of RPS on Key Dependent Variables by Political Viewpoint 67

7.4 Main Effects of Digital Game on Key Dependent Variables by Political Viewpoint 68

7.5 Effects of RPS on Different Types of Learning by Political Viewpoint 69

7.6 Effects of Digital Game on Different Types of Learning by Political Viewpoint 69

7.7 Summary and Comparison of Face-to-Face RPSs and Digital Games 70

8.1 Comparison of Face-to-Face RPSs and Digital Games 76

8.2 Human Health Effects of Climate Change 78

Appendix H Role-Play Simulation Workshop Sample Characteristics 103

Appendix L Digital Game Sample Characteristics 113

ACKNOWLEDGMENTS

This book is based on my dissertation research for a PhD in environmental planning and public policy from the Massachusetts Institute of Technology (MIT). Looking back, this book would not have been possible without the support and resources I received during my time in Cambridge.

In particular, I'd like to thank Lawrence Susskind for his exceptional mentorship. I'm also grateful to Jinhua Zhao for his advice on behavioral research and optimism about my dissertation, and Mariana Arcaya for her astute guidance on all matters methodological. I owe particular thanks to John Bolduc and Sam Lipson from the City of Cambridge for their support in testing the approaches and tools outlined in this book.

I would not have been able to successfully conduct this research without funding from the MIT Department of Urban Studies and Planning, the MIT Energy Initiative, the Martin Family Society of Fellows for Sustainability, and the Environmental Research and Education Foundation. Additional support was provided by a research seed grant from the Boston Area Research Initiative.

I owe many thanks to all the people who participated in this research, especially to the participants who gave their time to be interviewed. I hope I have done justice to the insights you shared with me.

I am grateful to Anthem Press for accepting this manuscript, and I'd like to thank Megan Grieving for her support during the manuscript preparation process.

Thank you to my family—my parents, my sister Michelle, and my brother Daniel—for their unwavering love and encouragement.

Finally, I am especially grateful to my husband and best friend Steven for inspiring me to write this book. Oppa, I couldn't have finished it without your homecooked meals and pep talks along the way, and I hope I can be as patient and supportive for your endeavors.

Chapter 1

INTRODUCTION

Recent years' increases in the intensity, frequency, and duration of hurricanes in the Atlantic and Pacific ocean basins, along with the incidence of the most intense (Categories 4 and 5) storms, are in part due to higher atmospheric and sea surface temperatures from climate change (Emanuel 2017; Melillo et al. 2014). More importantly, even if we were to stop emitting greenhouse gases today, with the lock-in of irreversible temperature increases from past emissions, climate impacts are expected to continue for the next few centuries (IPCC 2014; Melillo 2014; Solomon et al. 2009).

In 2017, Hurricane Harvey, a Category 4 storm, dumped more than 50 inches of rain over southeast Texas, resulting in more than eighty deaths and $180 billion of damage (Amadeo 2017). The historic and catastrophic devastation caused by the storm posed a host of public health concerns. Hurricane Harvey caused more than eight hundred wastewater treatment facilities and thirteen Superfund sites with toxic chemicals to flood (Hernandez, Zezima, and Achenbach 2017). Standing water in flooded homes in Houston was found to contain up to 135 times of what is considered safe of *E. coli*, a measure of fecal contamination (Kaplan and Healy 2017). News stories broke of how a senior citizen and a firefighter contracted necrotizing fasciitis, or "flesh-eating bacteria," from the contaminated floodwaters (Astor 2017). Long after the water drained and media attention subsided, post-Harvey residents had to grapple with increased risks of wound infections, diarrhea from toxic waters, and health threats from mold, mosquitoes, and other disease vectors, along with mental health stressors (Grigg 2017). The public health implications of climate change are significant.

As cities in the United States and around the world are increasingly experiencing the impacts of climate change, many are starting to include climate considerations in their planning and policymaking processes (Bierbaum et al. 2013; IPCC 2014). Cities are looking not only to mitigate greenhouse gas emissions from buildings, industries, and transportation to prevent future climate change but also to prepare for and manage climatic changes that have already been set in motion.

Urban climate adaptation to date has mostly focused on how cities can protect their physical assets from potential climate-related disasters, with an increasing emphasis on enhancing resilience, or creating places that can absorb and withstand climatic shocks (Hughes 2015). Scholars and practitioners have critiqued climate adaptation's current emphasis on building physical resilience to climate change, pointing out that adaptation plans rarely incorporate equity or social vulnerability (Hughes 2015). Consequently, calls have emerged for climate adaptation to focus on *human vulnerabilities* instead (IPCC 2014, ch. 15; Rumbach and Kudva 2011).

To that end, this book is about why and how the health impacts of climate change should be given a more prominent role in climate adaptation efforts at the local level. While the lack of attention to climate-related health risks in adaptation plans and policies have been pointed out by many, this has not yet led to climate adaptation planning and policymaking processes that situate citizens' health and well-being front and center. Therefore, cities will need new approaches to enhance awareness of and facilitate engagement with climate risk management choices that will build human resilience to climate change.

The *Frames and Games* Research Project

This book is based on the *Frames and Games* research project that was carried out from 2015 to 2017. The project was initiated with the expectation that an emphasis on the health impacts of climate change would increase people's understanding of and concern about climate change, and, furthermore, increase policy support for climate adaptation efforts at the local level. In addition, the project was designed to examine whether game-based public engagement approaches can be used to strengthen people's capacities to engage with climate adaptation planning and policymaking processes with a focus on health and well-being in their city.

The *Frames and Games* research project took place in Cambridge, Massachusetts—a city in the Northeast region known for its high educational attainment, progressive politics, and early climate adapter status within the country. A unique partnership between a research initiative at the Massachusetts Institute of Technology (MIT) and two municipal departments within the City of Cambridge enabled the *Frames and Games* project to collectively formulate a public health orientation to climate adaptation in municipal planning and policymaking processes.

Cambridge should be viewed as a friendlier place suited to attempting and testing the value of new tools in climate adaptation efforts at the local level rather than as a generalizable or representative example of other municipalities in the country. There are many lessons from designing and evaluating public engagement approaches in an early adapter city, particularly in enhancing awareness of technical issues and people's capacities to engage with policy choices that impact their health and well-being, that can inform climate planning and policymaking processes in other cities in the United States.

Introducing the Book

This book is intended to help readers understand how cities can reframe the conversation around climate change to be about public health. In addition, it aims to strengthen our understanding of different tools cities can use to operationalize a focus on the health implications of climate change, enhance collective decision-making capacities, and, ultimately, enable technically sound and politically feasible pathways to adapt to the changing climate.

The rest of this book is organized as follows. The first part provides the background for the research project featured in the book: Chapter 2 examines climate-health linkages and health-oriented climate adaptation frameworks, and Chapter 3 summarizes climate adaptation planning and policymaking efforts at the local level in the United States. The next part goes into more detail on the *Frames and Games* research project: Chapter 4 probes framing climate change as a public health issue, Chapter 5 examines the design and implementation of face-to-face role-play simulations, and Chapter 6 does the same for digital games. Finally, in the last part of the book, Chapter 7 compares the results of the project's two serious games, namely face-to-face role-play simulations and digital games, and Chapter 8 discusses the overall results of the *Frames and Games* project and its policy implications for policymakers and practitioners in furthering climate adaptation with a focus on human resilience to climate change.

Chapter 2

CITIES, CLIMATE CHANGE, AND PUBLIC HEALTH

Health Impacts of Climate Change

Over the last decade, deaths, injuries, and other health problems from floods, droughts, and other climate-related disasters cost an estimated $14 billion in the United States (IPCC 2014; USGRPC 2016; Watts et al., 2015, 1864). Climate change poses a series of significant and far-reaching threats to human health and well-being all over the world (IPCC, 2014; Kinney et al., 2015; Luber et al., 2014; USGCRP 2016), leading the World Health Organization (WHO) to declare climate change as the defining issue for public health in the twenty-first century (Chan 2006). In 2019, over seventy health organizations in the United States, including the American Medical Association and American Heart Association, issued a climate policy agenda to avert a public health crisis, calling climate change "one of the greatest threats to health America has ever faced" (Devitt 2019).

Focusing on climate-health linkages, Table 2.1 details how each climate impact translates into public health outcomes and indicates the populations most vulnerable to climate-sensitive health threats.

Planning for and Managing the Health Impacts of Climate Change

Climate change can affect health in two ways. As the climate continues to change, the risks to our health and well-being grow, due to either (1) an increase in the severity or frequency of health problems already affected by climate or weather factors or (2) unprecedented health problems in places where they did not previously occur (USGCRP 2016, ch. 1).

It is important to note that many climate-sensitive health outcomes are inherently place-specific. For instance, Stone's work on extreme heat documents how temperatures in cities can increase as much from local urban heat island effects as from global climate change impacts (Stone 2012). Sailor et al. (2016) find that this phenomenon tends to be underreported, as the IPCC's and other international and national climate assessments tend to focus on global average ambient air temperatures instead of locally experienced thermal stress, which is also dependent on surface temperatures, air pollution, humidity, and wind speeds (Sailor et al. 2016; Stone 2012).

Table 2.1 Human Health Effects of Climate Change

Climate Impacts	Health Impacts	Populations Most Affected
Extreme Heat	Premature death Cardiovascular stress and/or failure Heat-related illnesses, such as heat stroke, heat exhaustion, and kidney stones	Children Elderly Diabetics People with respiratory conditions People with cardiovascular conditions Outdoor workers Low-income populations
Poor Air Quality/ Pollution	Increased asthma, allergies, chronic obstructive pulmonary disease (COPD), and other cardiovascular and respiratory diseases	Children Elderly People with respiratory conditions People with cardiovascular conditions Outdoor workers Low-income populations
Wildfires	Injuries and death from burns and smoke inhalation Eye and respiratory illnesses from air pollution Exacerbation of asthma, allergies, chronic obstructive pulmonary disease (COPD), and other cardiovascular and respiratory diseases Population displacement, loss of home and livelihood	People with respiratory conditions
Extreme Weather Events	Injuries or death from drowning Lack of access to drinking water, water- and food-borne diseases from damage to potable water, wastewater, and irrigation systems Population displacement, loss of home and livelihood	Children Elderly Low-income populations
Increased Average Temperatures	Cardiovascular stress and/or failure Increased number and range of: Vector-borne diseases, such as West Nile virus, malaria Water-borne diseases, such as cholera, *E.coli* Food-borne diseases, such as *salmonella* poisoning Allergies from pollen	Children Elderly Diabetics People with respiratory conditions People with allergies Outdoor workers
All Impacts	Mental health effects, such as depression, anxiety, post-traumatic stress disorder (PTSD)	

Adapted from CDPH (2012), WHO (2013), Watts et al. (2015), and USGCRP (2016).

Locations with greater health threats from climate change include urban areas, particularly with heat island effects and air quality issues, coastal and other flood-prone areas, areas with higher levels of allergens and other air pollutants, and areas with limited access to water, energy, and transportation infrastructure (USGCRP 2016, 250). Consequently, the health impacts of climate change will vary across geographic regions and populations, making it imperative for local, state, and regional governments to prepare for the specific health outcomes their communities will face (Luber et al. 2014; Marinucci et al. 2014).

Current health-oriented climate adaptation planning and policymaking efforts exhibit three main shortcomings: First, they fail to fully incorporate or be structured around climate-health linkages, and subsequently, are not necessarily aligned with the end goal of alleviating the health risks of climate change. A majority of the climate plans that incorporate any public health elements merely acknowledge that climate change will have public health impacts, and fall short of delineating location-specific climate and health linkages and articulating localized strategies to alleviate the health risks of climate change. (CDPH 2012, 18)

This absence of a direct linkage between climate adaptation goals and actual human health outcomes leads to strategies that only indirectly address the underlying end goals of reducing negative climate impacts on health and improving overall health and well-being (Sailor et al. 2016). As the actual end goal of climate adaptation is, arguably, to alleviate threats to and improve the health and well-being of the city's inhabitants, instead of planting a certain number of trees or cooling the surface temperature of a city by a certain number of degrees, municipal policymakers and practitioners should situate climate-health interactions front and center in their climate planning efforts. For example, in contrast to tree-planting initiatives with metrics based on the number of trees planted or climate mitigation initiatives based on the number of degrees temperature increases should be confined to, the cities of Phoenix (HeatReady City program), Los Angeles (Urban Cooling Collaborative), and Philadelphia (Hot Weather-Health Watch/Warning System) are creating community cooling centers and neighborhood watch programs to reduce excess mortality and morbidity (Bierbaum et al. 2013; Union of Concerned Scientists 2019).

Operationally, climate adaptation planning and policymaking efforts should start setting goals based on health outcomes and improvements, and subsequently measure and evaluate policy progress toward these goals. Focusing on climate-health linkages can also contribute to a more robust planned adaptation to climate change, by (1) linking future climate impacts with more current policy concerns and assessing adaptation needs in a more inclusive way, (2) prioritizing co-benefits arising from adapting to climate change and reducing excess mortality and morbidity, and (3) reducing the risk of maladaptation actions arising from the uncertainty and complexity of climate projections, by focusing on low-regret or no-regret options, aligning to the principles laid out in Füssel (2007).

Second, the health impacts of climate change need further elaboration to enable understanding by both key stakeholders and the public. Research has shown that about two-thirds of Americans do not really give that much thought to how climate change might affect people's health. In addition, only about a quarter of the population is able to accurately name specific types of harm from climate change to human health (Maibach et al. 2015). The lack of knowledge of climate-health linkages and, specifically, about how climate change harms health outcomes today poses a substantial barrier to furthering public health-oriented climate adaptation (CPDH 2012).

Finally, climate adaptation plans and policies need to confront how the health impacts of climate change are uneven, not only across cities, but also within cities. Research has documented extensively how substantial health inequities that exist today (Arcaya et al. 2015) are only likely to be exacerbated by climate change (IPCC 2014; Patz et al. 2007; WHO 2013). People who live in places geographically vulnerable to climate risks tend to also lack the adaptive capacity to alleviate the health risks of climate change, making them disproportionately vulnerable to climate-related health risks. Without accounting for these differential impacts on different populations, climate adaptation plans and policies may fail to adequately reduce the health risks of climate change for those that are most at risk. For instance, current urban cooling strategies have been found to result in differential effects in reducing mortality rates according to age, income, and race (Vargo et al. 2016). In other words, unless health-oriented climate initiatives and programs engage with their publics in collective learning and dialogue, with a focus on expanding beyond the typical environmental and land-use planning actors and directly involving the people most vulnerable to these impacts, status-quo efforts are unlikely to address the health inequities of climate change (Sailor et al. 2016).

Two exemplars

In that sense, the Building Resilience Against Climatic Effects (BRACE) framework is a notable decision support tool in structuring climate adaptation planning and policy-making processes with the goal of reducing adverse health impacts of climate change. While the Centers for Disease Control and Prevention (CDC)'s BRACE framework (Building Resilience Against Climate Effects) was designed for local public health agencies to adapt to and manage the place-specific health impacts of climate change (Marinucci et al. 2014), the BRACE framework has much to offer to all municipal policymakers and practitioners aiming to reduce excess mortality and morbidity in cities facing climate change. BRACE defines (public health) adaptation as "any short- or long-term strategies that can reduce adverse health impacts or enhance resilience in response to observed or expected changes in climate and associated extremes" (Marinucci et al. 2014, 6436).

As many climate-related health threats are place-specific, BRACE aims to clarify and prioritize urgent climate-related health risks in a certain location. The iterative framework consists of (1) a vulnerability assessment, (2) modeling of projected health impacts, (3) an evaluation of intervention options, (4) development of a locally tailored adaptation plan, and (5) monitoring and evaluation of the plan (Marinucci et al. 2014). Table 2.2 provides more details on each of the five steps.

Through the Climate-Ready States and Cities Initiative, the CDC is assisting public health agencies in 16 states and two cities to apply the BRACE framework and prepare for the health impacts of climate change. The state grantees are Arizona, California, Florida,

Table 2.2 The Five Steps of BRACE: Building Resilience Against Climate Effects

Step 1: Anticipating Climate Impacts and Assessing Vulnerabilities Identify the scope of climate impacts, associated potential health outcomes, and populations and locations vulnerable to these health impacts.
Step 2: Projecting the Disease Burden Estimate or quantify the additional burden of health outcomes due to climate change.
Step 3: Assessing Public Health Interventions Identify the most suitable health interventions for the health impacts of greatest concern.
Step 4: Developing and Implementing a Climate and Health Adaptation Plan Develop a written plan that is regularly updated. Disseminate and oversee the implementation of the plan.
Step 5: Evaluating Impact and Improving Quality of Activities Evaluate the process. Determine the value of information attained and activities undertaken.

Adapted from Marinucci et al. (2014, 6436).

Illinois, Maine, Maryland, Massachusetts, Michigan, Minnesota, New Hampshire, New York, North Carolina, Oregon, Rhode Island, Vermont, and Wisconsin, and the city grantees are New York City and San Francisco. The grantees are in various stages of applying the BRACE framework. Most states have conducted vulnerability assessments, and about half have developed a climate and health adaptation plan. A notable few, such as Illinois, New Hampshire, and Oregon, have completed at least one iteration of the BRACE framework (APHA 2018).

Another noteworthy framework is the California Health and Equity's checklist for integrating public health into climate planning, shown in Table 2.3. This framework places more of an emphasis on the multi-stakeholder and multilevel governance aspects of climate adaptation for health with a focus on (1) building collaborative relationships between climate planning and public health agencies, (2) identifying community-relevant health goals and co-benefits of climate planning strategies, (3) specifying relevant climate and health indicators in the plan and subsequently measuring outcomes regularly, and (4) educating and engaging local policymakers, the media, community partners, and residents about climate and health linkages (CDPH 2012).

Conclusion

Current efforts to integrate the health impacts of climate change into climate adaptation plans and policies have three main shortcomings. First, location-specific climate-related health risks tend not be incorporated into climate adaptation plans and policies, and, subsequently, most climate adaptation plans are not aligned with the end goal of mitigating the adverse health impacts of climate change. Climate plans that incorporate any public health elements merely acknowledge that climate change will impact health and fall short of delineating location-specific climate and health linkages and articulating localized strategies to address these challenges. In addition, most climate plans do not address the lack of knowledge of climate-health linkages or better enable substantive

Table 2.3 Checklist for Integrating Health into Climate Planning

- Meet with local health department staff about climate planning process and implementation.
- Invite public health and other local health organizations to participate in climate plan development and coordinate and collaborate on implementation. Local health partners include hospitals, clinics, Lung Association local chapters, and community health organizations, many of whom are becoming more involved in land use, transportation, and climate and health issues.
- Make sure local policymakers understand the health and climate change connections and how these can be part of the overall climate plan.
- Identify and include health goals and co-benefits in a Request for Proposals for consultants, to ensure that health impacts will be integrated in the planning process early on.
- Identify relevant local health data and indicators for use in the climate plan.
- Identify public health co-benefits and potential adverse health consequences early in the screening, development, or implementation phases of the climate planning process. Health partners may be able to help with this analysis. For any identified negative consequences that may be associated with the climate plan, have a clear plan for mitigating or preventing these consequences.
- Identify health co-benefits that resonate most with the community's conditions and goals to ensure that the climate plan addresses the unique needs and interests of the local community.
- Include climate change and health information as part of community outreach and engagement during the development, adoption, and implementation phases.
- Identify health partners who can help with outreach, education, and communication strategies.
- As part of evaluating and reporting on climate planning progress, make sure that health outcomes are included and measured. Health partners may be able to help with this.
- When reporting progress to elected officials, media, partners, and residents, make sure to reinforce the human welfare, equity, and health benefits of measures to reduce greenhouse gases and strengthen community readiness.

Adapted from CDPH (2012, 13).

learning by both key stakeholders and the public. Finally, climate adaptation plans and policies need to confront how the health impacts are uneven not only across cities but also within cities and address the differential health impacts of climate change on different populations.

The BRACE and California Health and Equity's frameworks are two exemplars that address the outlined limitations of current efforts to integrate public health in climate adaptation plans and policies. The CDC's BRACE framework is an iterative framework that focuses on identifying place-specific climate-related health threats and designing a local climate and health adaptation plan to address these risks. The California Health and Equity's checklist for integrating public health into climate planning focuses more on building collaborative relationships between the relevant climate planning and public health agencies, reaching to the communities to identify existing health issues of concern that will be impacted by climate change, and engaging different stakeholders, including the media, to enable the publics to learn more about locally relevant climate-health linkages.

The *Frames and Games* project set out to address the outlined climate-health planning shortcomings and reflect the two exemplar frameworks in its design of the face-to-face and online public engagement tools. With an explicit focus on local climate-health linkages, these engagement tools can support cities prepare their publics to participate in collective decision-making processes to address the health risks of climate change. The serious games are based on an actual municipal vulnerability assessment with projected health impacts. In addition, the games induce the participants to focus on evaluating different climate policy options with the end goal of reducing excess mortality and morbidity. Finally, the simulated learning experiences guide the participants in envisioning different pathways to protect people's health and well-being against climate change.

Chapter 3

COLLECTIVE CLIMATE ADAPTATION AT THE LOCAL LEVEL

Planned Adaptation to Climate Change

Cities in the United States, and across the world, are beginning to take action to address increasing climate risks by creating adaptation plans. The Intergovernmental Panel on Climate Change (IPCC) defines climate adaptation as "the process of adjustment to actual or expected climate and its effects in order to either lessen or avoid harm or exploit beneficial opportunities" and climate adaptation planning as "the construction of actual strategies and plans in societies in response to climate change" (IPCC 2014, AG5 WGII, 871).

Numerous checklists and frameworks laying out the steps of climate adaptation planning have been put forth. The Fourth National Climate Assessment delineates five stages in the operational process of climate adaptation planning and policymaking: (1) awareness, (2) assessment, (3) planning, (4) implementation, and (5) monitoring and evaluation (USGCRP 2018, ch. 28). Susskind (2010) emphasizes the collaborative, iterative nature of determining when, whether, and how to respond to climate risks. After a municipality identifies problems likely to occur based on local climate forecasts and vulnerability assessments, key stakeholders in the city need to collectively decide what steps to take to reduce vulnerability, enhance resilience, and, finally, continuously monitor and revise adaptation plans in light of new information (Susskind 2010).

Climate adaptation planning is in its early stages in the United States. In 2014, the Third National Climate Assessment found many municipalities in the United States starting to assess vulnerabilities and develop standalone climate adaptation plans, but only a select few had initiated on-the-ground implementation (USGCRP 2014). The Fourth National Climate Assessment, published in 2018, found the scale and scope of adaptation activities to have increased (USGCRP 2018), but with less than 1 percent of municipalities in the United States undertaking climate adaptation planning overall, climate adaptation is still far from the norm (Hansen et al. 2013; Lyles et al. 2018). In short, while many cities have recognized the need for climate adaptation, a small number are planning for future climate impacts, and among the cities that have started the process of identifying local risks and creating strategies to deal with climate change, few are actually implementing adaptation plans (Shi et al. 2015; Woodruff and Stults 2016; Vogel et al. 2016).

Barriers to Climate Adaptation at the Local Level

There is a wealth of literature on factors that enable and constrain climate adaptation at the local level. Previous climate-related disaster experiences, perceived climate change threats, and political leadership have been cited as drivers of local climate adaptation action (Anguelovski and Carmin 2011; Betsill and Bulkeley 2007; Bierbaum et al. 2013). In addition, larger population sizes, higher levels of commitment from elected officials, and higher municipal expenditures per capita have also been shown to be associated with US cities pursuing adaptation planning (Shi et al. 2015). As Hughes (2015) notes, however, public support is rarely a driver for urban climate adaptation efforts. Instead, public opinion is often considered an impediment by cities in planning to adapt to climate change.

The barriers to local governments pursuing climate adaptation action are multifaceted and significant (Betsill and Bulkeley 2007; Bierbaum et al. 2013; Moser and Ekstrom 2010). Oft-cited barriers to climate adaptation planning and policymaking include (1) a lack of resources, such as financial resources or staffing capacity; (2) institutional barriers, including mismatched ecosystem and jurisdictional boundaries, a lack of legal mandates, political leadership, or operational control; (3) inherent difficulties in climate change decision making, due to the complexity in and uncertainty about future climate impacts, the lack of accessibility or utility of existing climate information; and (4) cognitive barriers, including limited awareness or apathy of decision-makers and the public; divergent risk perceptions, political beliefs, cultures, values, and limited integration of local knowledge and needs with technical information (see Bierbaum et al. 2013, 683–84, for a summary of adaptation barriers and references).

In particular, cities have found that most publics in the United States do not consider climate adaptation a political priority, even if they are aware of climate impacts and of its anthropogenic causes, highlighting the importance of conveying the necessity of climate adaptation to elected officials and other stakeholding publics (Carmin et al. 2012; Howe et al. 2015; Shi et al. 2015).

Framing the Policy Challenge

The framing of climate adaptation not only has been found to affect how climate adaptation is perceived relative to other municipal goals (Shaw et al. 2014) but also if and how climate adaptation plans are implemented (Dupuis and Knoepfel 2013). While the steps of assessing climate risks and identifying options to manage climate risks are presented separately in most climate adaptation planning frameworks, it is important to parse how a city defines a problem impacts its solution space (Rittel and Webber 1973). Accordingly, problem identification shapes how the city will go about enhancing the awareness of individuals, organizations, and institutions about climate change as well (IPCC 2014, ch. 16).

Urban climate change adaptation is often framed as protecting a city's physical assets from hazards and disasters (Hughes 2015). As a result, climate adaptation to date has focused on engineering protective structures to enable cities' built environments to better withstand climate-related events (IPCC 2014, ch. 15). For instance, Miami-Dade County's Capital Improvement Program is addressing hazards related to sea-level rise by raising the level of its roads and building seawalls to reduce flooding due to higher sea levels (USGCRP 2018). New York City is planning to build a series of levees, berms, and seawalls along the coastline of Staten Island to withstand a 300-year storm by 2022 (Kim 2019).

Consequently, with climate adaptation's current focus on protecting physical capital, critiques of climate adaptation have pointed out that adaptation plans rarely incorporate equity or social vulnerability (Hughes 2015), and calls have emerged for climate adaptation to focus on *human vulnerabilities* instead (IPCC 2014, ch. 15; Rumbach and Kudva 2011). The IPCC (2012) notes that how a climate vulnerability assessment starts, either by downscaling climate modeling outputs or assessing existing risks and vulnerabilities of the community of interest, results in fundamentally different problem orientations and solution spaces. Therefore, it is not coincidental that both the research and practice of climate adaptation tend to focus on identifying climate impacts and constructing defensive infrastructure to respond to the predicted climatic impacts, as is evident in a number of early adaptor cities' climate adaptation plans (Harries and Penning-Rowsell 2011; Hofstede 2008; IPCC 2012; Rosenzweig et al. 2011).

Collective Planning and Policymaking for Climate Change

Almost every local government in the United States with a climate adaptation plan has also committed to educating the public on the likely impacts of climate change and finding more productive ways to engage stakeholders in climate planning processes (Payne 2016; Hughes 2015; Tang et al. 2010). Regardless of the high proportion of cities tangibly making public engagement a facet of their adaptation policies, there have been continuous calls to further enhance the participation of stakeholders, from assessing risks and vulnerabilities to designing and implementing operational approaches in adapting to climate change (Hughes 2015; IPCC 2014; Moser and Pike 2015; Nordgren et al. 2016; Whitmarsh et al. 2013), implying that there is room for more and better public engagement in climate adaptation efforts. Indeed, a number of recent studies surveying local planners, managers, and officials involved with climate adaptation planning and policymaking processes show that building capacity in community engagement around climate change is a pressing need for most municipalities in the United States (Moser and Pike 2015; Nordgren et al. 2016).

While public engagement has always been a challenge for local governments in diverse policy domains, there are unique challenges to cities engaging with their publics and motivating action on climate adaptation, due to the magnitude, complexity, and uncertainties of climate change (Moser and Pike 2015; Susskind 2010). People tend to think that climate change will impact people and places that are geographically and temporally distant (Leiserowitz and Smith 2017). These inherent difficulties are only intensified by low levels of scientific literacy and education (Sterman 2011; Sterman and Sweeney 2007), concerted efforts at misinforming the public (Oreskes and Conway 2010), political polarization around the issue (Kahan et al. 2012; Kahan et al. 2013; Leiserowitz

2006), and lack of political leadership (Anguelovski and Carmin 2011; Bulkeley and Betsill 2005; Williams et al. 2017).

Aslin and Brown's (2004, 5) definition of public engagement is particularly relevant for the purposes of this book:

> Engagement goes further than participation and involvement. It involves capturing people's attention and focusing their efforts on the matter at hand – *the subject means something personally to someone who is engaged and is sufficiently important to demand their attention.* Engagement implies commitment to a process that has decisions and resulting actions. So it is possible that people may be consulted, participate and even be involved, but not be engaged. (Emphasis added)

In other words, public engagement in climate adaptation is the overarching process of involving the public in collective decision making around adapting to climate change, and the acts of consultation, participation, communication, and involvement are but each one method or means to that process (Moser and Pike 2015). In addition, an effective public engagement process would be one that results in a public that is "cognitively, emotionally, behaviorally, professionally, socially, spiritually, civically and/or politically involved and vested in the issues such that adaptation actions can advance" (Moser and Pike 2015, 112).

Why engage the public in climate adaptation? A variety of rationales support the call for public engagement in climate adaptation. In addressing the numerous barriers to climate change discussed above, many have argued that cultivating public and political support for climate adaptation is vital. Hughes explains, "Public opinion is not typically a driver of adaptation planning; instead, city governments are often working to build public support for adaptation planning and projects" (Hughes 2015, 24). According to the IPCC, deliberative public engagement processes help climate adaptation progress in cities (IPCC 2014, ch. 16). Enhancing public awareness of the impacts of climate change, engaging stakeholders in a city in identifying problems and solutions, and tailoring plans to fit the interests and needs of a community are critical in building public support for adaptation planning and implementation (Brunner and Nordgren 2012; Ligeti et al. 2007). While more effective public engagement alone is not enough to address all the barriers that cities face in local climate adaptation, better engagement is needed to overcome the numerous cognitive, cultural, political, and institutional barriers to climate adaptation outlined above.

In addition, as the IPCC's Special Report on Emissions Scenarios (2000) affirms, lifestyle and local development decisions will impact the climate more than any environmental policy, highlighting how the public conversation needs to go from a discussion of individual tweaking of behaviors or various technical solutions to a public conversation of alternative futures for the municipality, notwithstanding potential discussions of relocation and retreat (IPCC 2000, 2001). Susskind (2010) also points out the need for a collective discussion within a city to tackle the difficult questions of weighing tradeoffs to manage public risks. The long-term planning horizons and the uncertainty around future impacts of climate change make any anticipatory response to climate change controversial (Dessai and Hulme 2004; Tol 2003), especially when appropriate responses are potentially expensive or require a radical shift in the status quo (Few et al. 2007; Susskind et al. 2015).

Public engagement also facilitates community-based action and an overall more inclusive and deliberative approach to climate adaptation by integrating local knowledge, opinions, and needs into a technical-rational planning and policymaking process revolving around expertise and bureaucratic control (Few et al. 2007). Enhancing public engagement can be viewed as a way to focus on human vulnerabilities to climate change, as called for above, with a fuller consideration of equity and social vulnerability issues of the community of interest.

As can be seen, public engagement is promoted both principally as an end in itself, allowing communities to have more voice over decisions impacting their lives, and instrumentally as a means of collectively making better decisions (Few et al. 2007; Reed 2008; Rowe and Frewer 2000).

The proposed normative benefits of public engagement revolve around alleviating the marginalization of stakeholders on the periphery of a policymaking process by having more relevant stakeholders be included in a decision-making process impacting them (Martin and Sherington 1997). Participation can enhance trust in public sector decision-making processes (Rowe and Frewer 2000). Consultation and collaboration can empower stakeholders by co-generating knowledge and enhancing their capacity to utilize knowledge (Macnaghten & Jacobs, 1997; Reed 2008). Local knowledge and varied perspectives enhance problem-solving capacities (de Souza Briggs 2008). Public engagement also promotes social learning, where people learn from each other, resulting in increased trust and mutual understanding (Fung and Wright 2003; Innes and Booher 2004).

There are significant potential pragmatic benefits as well. Public engagement allows local governments and officials to identify the public preferences to incorporate in decision-making processes (Benhabib 1996; Fearon 1998), which can subsequently increase adoption and diffusion among target groups (Martin and Sherington 1997). Public engagement in environmental decision making can also lead to an increased sense of ownership among participants and may enhance the perceived equity and fairness of decisions, by accounting for diverse values and needs in human-environmental interactions (Reed 2008).

However, public engagement is not a solution to all planning and policymaking shortcomings, and has its limitations and drawbacks. Participation demands time and energy, and, indeed, the time and effort required to participate are the primary deterrents (Bloomfield et al. 2001). Consequently, the people willing and able to participate, and effectively bear the costs, tend to be disproportionately extreme view-holders (Fiorina 1999). Public engagement is constrained in addressing a fundamental lack of resources. For instance, in many science-intensive planning processes, community stakeholders have insufficient expertise to engage in technical debates (Fischer and Young 2007). Solutions derived from a public engagement process may not be feasible (Day 1997; Fiorina 1999; Thompson 2005). Participation can reinforce existing inequalities (Kothari et al. 2001), and groupthink may prevent less popular perspectives from being considered (Nelson and Wright 1995).

More importantly, people may not want to participate because of past relationships and previous negative experiences (Potter 1985). Consultation fatigue is another common phenomenon (Duane 1999; Wondolleck and Yaffee 2000). In addition, self-confidence and respect for authority are known to influence people's decisions on whether to participate (Sanchez et al. 1988).

Many typologies of public engagement have emerged, starting from Arnstein's (1969) seminal "ladder of citizen participation," with rungs ranging from manipulation to informing to consultation to partnership to citizen control, based on the degree to which stakeholders are engaged and the role of the public in decision-making processes. White (1996) presents four types of participation based on the extent of participation among stakeholders at the top and grassroots movements at the bottom: nominal, instrumental, representative, and transformative, also based on the normative assumption that empowering individuals and organizations at the bottom results in more genuine forms of public engagement. Susskind (1987; and with Field [1996]) presents a participatory and collaborative approach to public policymaking built around the mutual gains approach to negotiation, consisting of the four steps of preparation, value creation, value distribution, and follow-through (Carson 2008; Susskind 1987; Susskind and Field 1996).

However, according to Few et al. (2007), public engagement in climate policy is particularly difficult with a fundamental tension between strategic, long-term planning and meaningful inclusion. They assert that climate planning cannot be a "bottom-up" process of decision making, as the most fundamental decision to respond to climate change has already been made. Few et al. (2007) caution that narrower approaches to participation that solidify stakeholders' commitment to implementing climate adaptation plans instead are more likely to succeed, as long as the scope and limitations of public engagement in this context are explicitly communicated from the beginning of the process. On the more extreme end, Renn (2006) argues for an even narrower conception of public engagement in climate planning and policymaking processes, with expert-led discussions and the public checking the value judgments of experts.

Rowe and Frewer (2000) focus on the direction of information flows in interactions among stakeholders in their typology of public engagement: communication (top to bottom, one-sided), consultation (bottom to top, one-sided), and participation (two-sided). Farrington and Bebbington (1993) use an axis to map out different forms of participation, with depth as one axis, indicating degrees of involvement, and breadth, representing inclusion and exclusion, as the other axis, indicating the tradeoff that many public engagement efforts face in involving the public in different planning processes. For example, face-to-face public engagement efforts are high on the depth axis and may have more potential for transformative learning (Rumore et al. 2016; Susskind et al. 2015) but may find it difficult to secure broad-based participation beyond the "usual suspects" and score equally high on the inclusion axis (Beierle 1999; Konisky 2001). In contrast, public engagement using digital technologies pose lower costs and barriers to entry to participants, indicating more potential for breadth, but questions remain on the level of depth possible with online forms of public engagement (Best and Krueger 2005; Evans-Cowley and Hollander 2010; Gordon and Baldwin-Philippi 2014).

There has also been an extensive discussion of choosing the appropriate method of engagement. Arnstein (1969), Biggs (1989), and Pretty (1995) argue for using different methods for the different rungs of their ladder of participation. Rowe and Frewer (2000) present a range of public engagement methods based on the direction of information flows. When communicating with the public, information can be disseminated through mass media, pamphlets, hotlines, and public meetings. For consulting the public, consultation documents, opinion polls and referendums, focus groups, and surveys can be utilized. For participation, they discuss using citizens' juries, consensus conferences, task forces, and public meetings with voting (Rowe and Frewer 2000). Susskind (2010) presents various tools for creatively engaging stakeholders in climate adaptation planning and policymaking processes: conflict assessment with an emphasis on fostering readiness to participate in collaborative efforts and not just understanding the technical issues, joint fact-finding, scenario planning, and role-play simulations. Along with findings indicating that different tools are appropriate for different contexts, research also indicates that there is no one preferred tool for public engagement and that public engagement is best supported through the use of a range of participative tools (Carson 2008; Few et al. 2007).

How would we evaluate the outcomes of public engagement in climate adaptation planning and policymaking processes? Beierle (1999) argues for using a set of social goals to evaluate the outcomes of public participation in environmental decisions: (1) educating the public; (2) incorporating public values, assumptions, and preferences into decision making; (3) increasing the substantive quality of decisions; (4) fostering trust in institutions; (5) reducing conflict; and (6) making decisions cost-effectively. In contrast, Mandarano (2008) argues for a consideration of both social and environmental outcomes in evaluating engagement in collaborative environmental planning efforts. She presents a framework weighing (1) outputs (high-quality documents, such as plans and collaborative science), (2) social outcomes (increases in social, intellectual, and/or political capital, innovation, and institutional change), and (3) environmental outcomes (changes in environmental parameters, such as land cover or water quality, and perceptions of environmental quality).

Conclusion

Taken together, the wealth of literature on public engagement stresses the importance of public engagement as a key feature of climate planning and policymaking efforts, but research on how to go about engaging the public in the climate adaptation processes is relatively limited. Moser and Pike summarize the status quo succinctly: "Current local adaptation efforts are thus faced with (1) a growing and persistent need for effective communication and public engagement, (2) a pervasive lack of capacity to do so, and (3) limited opportunities to date for building that capacity" (Moser and Pike 2015, 112). Hughes also highlights this gap both in practice and research: "We require a better understanding of the role that the public is and should be playing in urban climate change adaptation planning and how this role affects capacity for action" (Hughes 2015, 24).

The *Frames and Games* research project was designed in response to Hughes's (2015) and Moser and Pike's (2015) call for evidence-based research in exploring how city governments can best engage the public in adaptation planning, by exploring how a public health orientation to climate change can be utilized to engage with different publics and foster collective learning in climate adaptation planning and policymaking processes.

Chapter 4

FRAMING CLIMATE CHANGE AS A PUBLIC HEALTH ISSUE

Issue Framing and Public Opinion

The framing of a policy issue has been called the "essence of public opinion formation" (Chong 1993, 870). Indeed, framing is one of the primary means policymakers can use to influence public opinion; the way an issue is framed matters for how publics think about it (Sniderman and Theriault 2004). As Entman summarizes the phenomenon, "To frame is to select some aspects of a perceived reality and make them more salient in a communicating text, in such a way as to promote a particular problem definition, causal interpretation, moral evaluation, and/or treatment recommendation for the item described" (Entman 1993, 52). In other words, a framing effect occurs when in describing an issue or event, "emphasizing a subset of potentially relevant considerations causes individuals to focus on these considerations when constructing their opinions" (Druckman 2001b, 230).

For example, framing the discussion about a Ku Klux Klan rally as a public safety or free speech issue shapes public opinion in widely diverging ways (Druckman 2001b). Similarly, even with the same details, public acceptance of a potential policy to control a rapidly spreading disease is higher when the policy solution is framed as saving a number of lives compared to when the policy is framed as losing others (Tversky and Kahneman 1981). Framing effects are understood as a result of heuristic information processing (Kühberger 1997; McElroy and Seta 2003), with frames serving as cognitive shortcuts to help people make sense of complex information (Kaufman et al. 2003).

There are two types of framing effects. The first, the equivalency framing effect, results in people modifying their preferences when "different, but logically equivalent" words or phrases are used to describe the same phenomenon (Druckman 2001b, 228). For instance, people view the same glass of water differently depending on whether the glass is described as half-full or half-empty. In the previously mentioned seminal study of the equivalency framing effect, the Asian disease task paper, Tversky and Kahneman (1981) elicited different policy preferences from people by framing the same problem as either an issue of saving lives or minimizing expected deaths, showing that the framing of questions as losses or gains systemically reverses people's choices.[1] The cognitive

---

1 Participants in the positive frame condition received a description of lives saved (e.g., 200 out of a total of 600 people will be saved), whereas participants in the negative frame group were informed about lives lost (e.g., 400 out of a total of 600 people will die) in making a decision.

psychology and behavioral science fields have studied the equivalency framing effect extensively. They have found that frames that cast "the same critical information in either a *positive* or a *negative* light" cause people to have different preferences (Levin et al. 1998, 150; emphasis in original). For instance, people evaluate ground beef as better tasting when it is labeled as 75 percent lean, rather than 25 percent fat, even though the substance of the beef is equivalent, regardless of the description (Levin and Gaeth 1988).

The second type of framing effect, known as the emphasis framing effect, occurs when "emphasizing a subset of potentially relevant considerations causes individuals to focus on these considerations when constructing their opinions" (Druckman 2001b, 230). For example, when government spending for the poor is framed as increasing the chances of the poor leading fuller lives, people tend to support increased spending. However, when government spending is framed as leading to higher taxes, people will oppose the same increase in government spending. While the issue is the same, Sniderman and Theriault (2004) explain that the "enlarging opportunity" and the "higher taxes" frames appeal to different audiences depending on their underlying political principles.

Nelson et al. (1997) are careful to distinguish this type of framing—emphasis framing—from priming, or merely presenting new information. Priming, through making particular considerations more accessible temporarily, influence spur-of-the-moment decision processes. Unlike priming, frames tell people how to weigh existing information differently, by placing more emphasis on the particular considerations made more salient through a specific frame of the issue. In summary, attitudes regarding an issue can be shifted by (1) adding new information to one's considerations about the issue (belief change), (2) making particular considerations temporarily more accessible (priming), or (3) altering the weight of particular considerations (framing) (Nelson et al. 1997, 236). It is the last method of framing, emphasis framing, or encouraging people to attach greater weight to particular considerations of climate change, which the *Frames and Games* research project sought to probe in more detail.

Frames and Climate Change

In light of the formidable barriers to improving public understanding of the sources and impacts of climate change, Nisbet (2009) advocates for using carefully tailored frames that take account of existing attitudes and values, and, furthermore, amplify the personal relevance of climate change for different publics. To better communicate about and frame climate change as a policy issue, Moser and Dilling (2004), Fischhoff (2007), and Weber and Stern (2011) recommend harnessing the impact of information framing, anchors, and heightening the availability and salience of examples by focusing on people's deepest concerns in climate communication. Nisbet and Scheufele agree: "Effective communication will necessitate connecting a scientific topic to something the public already values

---

Despite the equivalency of the two versions, people are more likely to be risk-averse when outcomes are framed as gains and more risk-seeking when outcomes are framed as losses.

or prioritizes, conveying personal relevance. And in people's minds, these links are critical for making sense of scientific information" (Nisbet and Scheufele 2009, 1774).

Addressing concerns over potential manipulation of public opinion, Nisbet and Scheufele state that "framing should not be seen as a strategy for 'selling' the public on science, but rather as a means for constructively shifting the conversation about an issue" (Nisbet and Scheufele 2009, 1771). Empirical research has also identified numerous factors that decrease framing effects, indicating that framing is far from a one-size-fits-all policy tool that induces all receivers of a certain frame of an issue to passively accept its influence. Heterogeneity in framing effects can be explained by individual characteristics, such as the need to evaluate (Druckman 2001b; Smith and Levin 1996), and the credibility of the frame's source, or how knowledgeable and trustworthy the source seems (Druckman 2001a). In addition, after being exposed to a certain frame, group deliberation of an issue with people holding divergent views (Druckman and Nelson 2003) and certain cognitive tasks, such as risk assessments and cost-benefit analysis, reduce framing effects (Hodgkinson et al. 2002; Sunstein 2000; Wright and Goodwin 2002). In other words, while framing a policy issue can be effective in highlighting a certain problem orientation and enhancing awareness of the issue, as framing relies on heuristic decision making, the efficacy and longevity of the frame depends on whether and how the receiver of the frame also engages in more deliberate, effortful thinking around the policy issue.

Research to date on framing the policy issue of climate change has tended to focus on testing the effects of equivalent framing by highlighting either the positive or negative attributes of climate mitigation efforts and its effects on attitudes around climate change. Studies have found that positive frames that highlight gains (in terms of avoided negative impacts) resulted in more positive attitudes toward climate mitigation actions than negative frames that highlight losses (in terms of likely climate impacts without mitigation) (Morton et al. 2011; Spence and Pidgeon 2010). In addition, messages that frame the collective benefits of climate action, rather than highlighting costs in terms of necessary sacrifices, were found to result in higher perceived capacity to deal with climate change and stronger behavioral intentions regarding climate mitigation (Gifford and Comeau 2011). However, studies testing the emphasis framing effects of climate change, by highlighting *different* implications of climate change, are relatively rare.

Framing Climate Change as a Public Health Issue

A public health framing of climate change, which is central to the *Frames and Games* research project, is a potential way of fostering collective awareness of and increasing public engagement with the issue. This frame emphasizes the potential of climate change to increase the occurrence of health problems that are already perceived as important, such as infectious diseases, asthma, allergies, and heat stroke (USGCRP 2016). A public health frame also tends to heighten the geographic proximity of impacts by substituting conceptions of remote regions and abstract populations with one's own neighbors and places. In the process of making climate change easier to understand, a public health framing can also increase its salience, or the relevance of the information, and its legitimacy, or the consideration of appropriate values and concerns (Cash et al. 2002; Frumkin et al. 2008; Maibach et al. 2010; Nisbet 2009).

There is a research gap, however, on how a public health framing of climate change risks impacts perceived knowledge of and concern for climate change and subsequently affects policy support for climate adaptation efforts (Leiserowitz et al. 2013; Maibach et al. 2010; Nisbet 2009). Can a public health frame of climate change contribute to alleviating the social and institutional barriers to climate adaptation at the local level?[2] As seen earlier, previous research in the framing of climate change as a policy issue has tended to focus on the effects of equivalent framing (positive versus negative frames; benefits versus costs) on knowledge of and engagement with climate mitigation activities.

As an exception, Myers et al. (2012) studies the emphasis framing effect by testing the effects of considering the environmental, public health, or national security implications of climate change and how they elicit different emotions regarding climate change. They found the public health frame to be more effective than the environment frame in eliciting hope, while the national security frame actually elicited anger. As they focused on emotional outcomes, they did not examine whether focusing on public health impacts of climate change might make it more understandable, salient, or personally relevant. They also did not study how frames of climate change might affect public concern about the policy and collective commitment to climate adaptation planning and policymaking efforts in their own municipality. In summary, there is yet to be a study that analyzes the emphasis framing effects of different frames of climate change on local climate adaptation efforts.

In response, the *Frames and Games* project designed and implemented a study to evaluate the influence of two ways of framing climate change risks (i.e., in terms of environmental impacts or public health impacts) by discerning how each framing of the issue of climate change affects perceived levels of knowledge about climate change, the perceived severity of the threats posed, concern for local climate risks, confidence in the capacity of local governments to undertake effective climate adaptation action, willingness to pay higher taxes for local climate adaptation, and policy preferences regarding climate change at the local level. This emphasis framing study was initiated with the main hypothesis that an emphasis on the health impacts of climate change would increase people's understanding of and concern for climate change and, furthermore, increase policy support for climate adaptation and policymaking efforts.

Research Process

The study employed a randomized, controlled message experiment, where participants were randomly assigned to one of two experimental conditions. Participants read one of two differently framed vignettes about climate change, each approximately 300 words, emphasizing either the risks to the environment or the risks to public health. Other than the environment and health considerations being highlighted, the messages were structured identically. The vignettes used in the survey are shortened versions of the environment and health framing articles used by Myers et al. (2012). The two vignettes are presented in Appendix A.

---

[2] See Chapter 3 for a summary and discussion of barriers to climate adaptation at the local level.

After reading the randomly assigned vignette of either the environment or health frame of climate change, participants answered a series of questions about possible responses to climate change. In particular, the following questions were posed to participants in both groups to gauge cognition, affect, and behavioral intentions[3] regarding local climate adaptation in their municipality:

a) perceived knowledge of climate change[4] [cognition]
b) perceived seriousness of threats from climate change[5] [affect]
c) concern for climate impacts in their municipality[6] [affect]
d) confidence in effective local climate adaptation action[7] [affect]
e) support for city incorporating climate risks into everyday planning[8] [behavior]
f) willingness to pay higher taxes for their city to reduce climate risks[9] [behavior]

The survey questions are in Appendix B.

Expected results

Compared to the participants that were exposed to an environment frame of climate change, participants exposed to a health frame of climate change were expected to have, on average:

a) a higher level of perceived knowledge of climate change [cognition]
b) a higher level of perceived seriousness of threats from climate change [affect]
c) a higher level of concern for climate impacts in their municipality [affect]
d) a higher level of confidence in effective local climate adaptation action [affect]
e) a higher level of support for city incorporating climate risks into everyday planning [behavior]
f) a higher level of willingness to pay higher local taxes for their city to reduce climate risks [behavior]

Results

A total of 169 people participated in the online survey experiment between September 2015 and May 2016. The participants were randomly assigned to two groups and were

---

3 In social psychology, attitude is defined as "an individual's disposition to respond favorably or unfavorably to an object, person, institution, or event, or to any other discriminable aspect of the individual's world" (Ajzen 1989, 241). Rosenberg and Hovland (1960) posit in their tripartite model of attitude structure that one's attitude can be inferred by measuring three components of cognition, affect, and behavior. This tripartite model of attitude structure is the most prevalent conceptual model of attitude in social psychology research (see Ajzen 1989; Allport 1964; Hilgard 1980; McGuire 1969 for a review). See Bagozzi et al. (1979), Breckler (1984), and Eagly and Chaiken (2007) for a discussion regarding the construct validity of the tripartite classification of attitudes.

4 Validated and tested questions from Lee et al. (2015).

5 Validated and tested questions from Lee et al. (2015).

6 Validated and tested questions from Rumore et al. (2016).

7 Validated and tested questions from Rumore et al. (2016).

8 Validated and tested questions from Rumore et al. (2016).

9 Question adapted from the General Social Survey (GSS) on willingness to pay higher taxes for environmental protection.

Table 4.1 Main Effects of Framing on Key Dependent Variables

Dependent Variables (n = 169)	Framing Effect Δ Health-Env	Cohen's *d*
Perceived knowledge of climate change (Likert scale: 1–3)	**0.148**** (0.076)	0.303
Perceived personal threat from climate change (Likert scale: 1–5)	0.005 (0.118)	0.006
Concern for local climate risks (Likert scale: 1–5)	**0.351**** (0.164)	0.33
Confidence in city's effectiveness in climate adaptation[a] (Likert scale: 1–5)	**0.509***** (0.104)	0.629
Support for city incorporating climate risks into everyday planning (Likert scale: 1–5)	0.009 (0.096)	0.014
Willingness to pay higher taxes for climate adaptation (Likert scale:1–5)	−0.015 (0.148)	0.015

Note: Standard errors in parentheses. All mean comparisons significant at *p < 0.5 (bold face), **p < 0.1 (bold face), ***p < 0.01 (bold face). Unequal variances assumed (t-test). Cohen's *d* is a standardized measure of effect size; values between 0.3 and 0.6 are considered "moderate" effect sizes in behavioral sciences (Cohen 1988).

[a] n = 56

exposed to either the environment frame or the health frame. The environment frame group consisted of 85 randomly assigned participants, and the health frame group consisted of 84 participants.

A series of independent t-tests[10] were conducted to determine if there were differences in their knowledge, affect, or behavioral intentions regarding local climate adaptation based on the frame of climate change they were exposed to (i.e., the environment frame or the public health frame).

Table 4.1 sums up the main effects of framing climate change as either a public health issue or an environmental issue. In line with expectations, participants exposed to a health frame showed statistically significant higher levels of perceived knowledge of climate change, concern for local climate risks, and confidence in their municipality's capacity to effectively plan for future climate impacts when compared to the participants exposed to the environment frame of climate change. Participants exposed to the health frame did not show statistically significantly higher levels of perceived personal threats

[10] This study acknowledges the debate around using parametric tests on Likert scale data. While principled stances such as Jamieson (2004, 1217) forbid conducting parametric analyses on ordinal data as "intervals between values cannot be presumed equal," empirical studies of the robustness of tests of central tendency, including t-tests and ANOVA, find that parametric tests on differences in means for sample sizes greater than 5 show robustness with respect to ordinality and non-normality (Boneau 1960; Norman 2010; Pearson 1931). In addition, Wilcoxon signed-rank tests were also conducted with similar results, enhancing confidence in the parametric test outcomes.

from climate change, support for their city incorporating climate risks into everyday planning, or willingness to pay higher taxes for their municipality to reduce climate risks than people exposed to the environment frame of climate change.

In other words, framing climate change as a public health issue evokes a stronger sense of concern regarding climate risks and increases public commitment to local climate adaptation action when compared to depicting climate change as an environmental issue. With its emphasis on the projected increases in existing health issues and the geographic and social proximity of the likely impacts of climate change, a public health frame of climate change can make climate change seem easier to understand and address at the local level. This indicates that cities have much to gain from framing climate change as a public health issue rather than as an environmental issue, as the health and well-being emphasis can boost public concern about the severity of the problem and raise public support for policy action.

Thoughts on the efficacy of the public health frame

About twenty semi-structured interviews were conducted with survey experiment participants to examine people's thoughts on using a public health frame of climate change to enhance awareness of and foster engagement with climate change. According to the interview results, the public health frame heightens the geographic, temporal, and social proximity of climate impacts and amplifies people's interest in *climate adaptation at the local level*, corroborating previous research on the efficacy of the health frame of climate change as a public education and engagement tool for *climate mitigation* activities. The public health frame of climate change also draws attention to climate change's potential to increase the prevalence of infectious diseases and other health problems that are already perceived as important, especially among the most vulnerable populations in a city or region. Most of the interviewees thought that climate change should be framed as a public health challenge, but some had reservations about using this frame to reach different publics.

People supported the use of the health framing of climate change or an emphasis on the health implications of climate change for different reasons. One participant advocated for additional information and engagement initiatives focused on the health impacts of climate change from a right-to-know principle: "I know it may seem alarmist to some people, who are not open to hearing about the health risks that climate change poses, but I think everyone has the right to be aware of health impacts that may threaten your health, and individuals are best equipped to take care of themselves or take actions to lessen the risks."

Others pointed to the potential of the health frame of climate change to resonate with people in terms of what they already think is important in their lives, instead of imposing climate change as an important policy issue per se. One participant working with vulnerable populations argued, "When you go to communities and start talking about the things that concern them, issues like asthma resonate. That affects people very tangibly. It affects more low-income, people of color, poor housing conditions, etc. By starting with what concerns people, and helping to address it immediately, you can also broaden into other areas."

Finally, many participants interviewed pointed to how they thought the health frame of climate change could politically motivate fellow citizens on climate adaptation. In particular, the interviewees noted that despite public engagement efforts to communicate climate risks to the public, a substantial portion of the public (about 30 percent, according to Leiserowitz et al. 2013) remain disengaged or dismissive about climate change. In this landscape, one interviewee specified the health frame's efficacy in transcending the skepticism around and fossilized knee-jerk reactions to climate change: "I'm very hopeful that people are willing to talk about climate change in ways that seem to disengage the 'shields' that have been built around it in the last 8–10 years, where some people in certain fields just don't want to listen to anything about it anymore. But I think health is a way to maybe get around that."

This interviewee went on to explain that he realized other people were just as interested in climate change for health and well-being reasons as well:

> I was encouraged that there are other people saying, "The reason I'm interested in climate change is because I'm interested in the health of my children." And I'm not about climate change as my first goal. My first goal is a healthy environment for my children and for the animals on the planet, for the whole biosphere. And that's really what we're fighting for. For health.

One interviewee talked about an analogous experience in successfully using a public health frame of climate change for a campaign to reduce idling in cars and trucks:

> In terms of public health, I thought of asthma. I would go to the drivers and say, "I'm concerned about asthma in our kids" and people would say "Oh thank you" and turn off their engines. If I said anything about climate change, the response would be much worse. The more we can put a face to all this stuff, the more powerful it is.

She went on to say,

> I find that talking about climate change in terms of health is a new idea to most people. When people hear about the co-benefits of reducing carbon emissions, they tend to roll their eyes because they think we're talking about benefits occurring one hundred years from now, or how it'll affect people in Bangladesh, and people think "Well, I have problems here today." So we talk about how these activities impact your health today, the air you breathe today, and the water you drink today.

However, there were also some people that were skeptical or critical of the public health frame. A couple of people interviewed wanted to highlight the risks of climate impacts more directly. One interviewee remarked, "New terms are needed. Public health is too clinical. Neighborhoods don't even talk about rats, even though no one likes them. Discussing public health doesn't help. We need to get to the specifics [of climate change]." Similarly, another interviewee argued, "'Public health' is a weasel word used to not upset the populace. Don't use that! Say what is real! This stuff is non-partisan if you help people understand what is involved and what will happen, in plain English.

That will really have an impact." Apart from the substantive health implications of climate change, the phrase "public health" itself and the field of public health seems to conjure certain images or sentiments to people, influencing how they perceive the efficacy of using a public health frame of climate change to engage different publics around the policy issues.

Others felt that a public health frame would cause fear and disengagement. One interviewee explained that she intentionally chooses to discuss climate change with a select audience of like-minded friends or colleagues because she is afraid of scaring people away: "Doing things that make us fearful isn't always helpful either [...] I only really talk about [health impacts of climate change] with my climate buddies because most people don't really want to talk about it. It is so big, it is so devastating." Another interviewee emphasized that a public health frame needed to be coupled with what governments are doing to alleviate the health risks; otherwise, it would be better not to present just the dangers or the risks at all: "We need to keep stakeholders informed, but also with the assertion that the city is proactive and trying to keep ahead of these developments." These comments point to the importance of not just attracting more attention to a policy issue with the use of certain frames but also establishing self-agency by explaining what people can do with the information. To this effect, climate communication research has shown that people sustain knowledge of climate risks better when information is coupled with steps they can take to reduce vulnerabilities and prepare for climate-related events (Pidgeon and Fischhoff 2011).

Finally, others felt that a frame of economic losses or property damage would be more effective. One interviewee asserted, "A better [framing] would be in terms of loss of property values. Flooded property becomes zero in value, a zero tax base. People will move away. If you can't sustain your city, then you don't have a city." Indeed, while the *Frames and Games* project focused on framing climate change as a public health issue (versus an environmental issue), other potential frames of climate change, such as economic development or national security, may be also effective in amplifying climate change as a policy issue to certain publics. While past research on climate change and frames has focused on testing the equivalency framing effects of highlighting the economic benefits of climate mitigation activities, testing the efficacy of other potential frames of climate change, such as economic development or national security, against the health framing of climate change in boosting public concern of and commitment to municipal action warrants further research.

Conclusion

To address the research gap on how different frames of climate change affects attitudes around climate adaptation, the *Frames and Games* research project implemented a study to test the influence of framing climate change risks (i.e., in terms of environmental impacts or public health impacts) on knowledge about climate change, concern for local climate risks, and policy support for climate adaptation action. While previous research in framing and climate change has tended to focus on the effects of equivalent framing (positive versus negative frames) on climate mitigation, this study is the first to analyze the emphasis framing effects of climate change on local climate adaptation efforts.

Overall, the results of this study provide support for the main expectation of the *Frames and Games* project. Focusing on the health impacts of climate change increased perceived knowledge of climate change, concern for local climate risks, and confidence in one's local government's capacity to effectively plan for climate risks.

These results provide proof-of-concept that a public health frame of climate risks, as compared to an environment frame, increases concern for climate change and support for greater investment in climate adaptation at the local level, implying that cities have much to gain from framing climate change as a public health issue. The next two chapters will delve into face-to-face and digital games that cities can use to operationalize the public health framing of climate change that has been tested in this study and, furthermore, enable people to collectively devise adaptation pathways to enhance the human resilience of their city against climate change.

Chapter 5

ROLE-PLAY SIMULATIONS

Serious Games and Climate Change

Social barriers to climate adaptation primarily operate at the individual and collective decision-making levels (Hulme et al. 2007). For instance, at the individual level, a perceived lack of self-efficacy may hinder some from preparing for climate change. At the collective level, climate adaptation planning and policymaking tends to be characterized by a status quo bias, often catalyzed by the lack of perceived collective efficacy (Bandura 1998; Hulme et al. 2007). In response to these deep-seated personal and collective barriers to climate adaptation, Rumore et al. (2016) argue for the need to enhance communities' readiness to adapt to climate change. In their view, acquiring substantive knowledge, or what they call adaptation literacy, at both the individual and community level, is necessary but not sufficient for a community to achieve climate readiness. Collaborative capacity, or the ability to work together to solve common problems, is key (Rumore et al. 2016). A ready community would possess a "shared sense of the risks that communities collectively face, how they might prepare for and manage aforementioned risks, and different decision-making approaches to respond collaboratively and adaptively to emerging threats" (Rumore et al. 2016, 746). However, current public engagement practices in planning are likely to be inadequate in their collective technical and political learning processes for climate-ready planning (Argyris and Schon 1978; de Suarez et al. 2012; Evans-Cowley and Hollander 2010; Susskind 2001).

How can we accelerate individual and collective learning to advance readiness in climate adaptation? Based on decades of research, we know that learning experiences are most effective when they are active, experiential, and have shorter feedback loops (Connolly et al. 2012; Kolb 2014). Accordingly, serious games, or games with an explicit purpose of learning complex material or decision-making methods through play, have been gaining attention as a potentially powerful tool for transformative learning and engagement (Abt 1987; de Suarez et al. 2012; Gee 2003; Shaffer 2006; Squire 2003, 2006).

As experiential exercises presenting fictional challenges within clear constraints (Abt 1987), serious games can be used to induce people to take a long-term perspective, to use systems thinking, and to experiment with problem-solving strategies (McGonigal 2011). Serious games have also been found to be effective in conveying complex information and enabling perspective-taking (Abt 1987). These games have been used to attempt to shift behaviors and attitudes in various areas of public policy, including health, environmental policy, and climate change (de Suarez et al. 2012; Dolin and Susskind 1992; Rumore et al. 2016; Sawyer and Smith 2008; Wu and Lee 2015). Far from being a recent development, however, games have been utilized in planning throughout history (Light 2008), including the Model Cities Program (Berkeley 1968) and more recent game-based approaches to participatory planning (Gordon and Baldwin-Philippi 2014).

Role-play simulations are a type of serious game where participants engage in a face-to-face mock decision-making process structured by instructions delineating role-specific interests and strategies. Each participant receives two sets of instructions: (1) general instructions with the background of the policymaking process in the context of the locality and (2) role-specific instructions with more details on the interests and positions of the role the player will assume for the game (Rumore et al. 2016). Role-play simulations offer interactive ways of mastering technical information within fixed rules and constrained outcomes. In addition, role-play simulations have been shown to help participants learn not only technical knowledge but also process-based knowledge, such as existing social norms and institutional arrangements within a policy realm (Rumore et al. 2016).

Role-play simulations have been highlighted as particularly promising tools for education and public engagement. First, role-play simulations immerse people in scenarios where a sense of "here and now" is crucial, such as medical, negotiation, and military training, and enable them to learn to deal with high-stakes circumstances in a safe and low-cost sandbox environment (Rumore et al. 2016; Schenk 2015). Well-designed games can prepare people for critical situations where certain decisions can result in life or death outcomes in the real world, like that of dealing with climate change (de Suarez et al. 2012).

In addition, in the planning and policymaking context, role-play simulations help people understand different stakeholders' viewpoints and interests regarding the policy realm featured in the game. By forcing participants to take on a role different from the ones they occupy in real life, role-play simulations foster perspective-taking and empathy for others' perspectives, and subsequently can increase participants' capacity to collectively problem-solve in their communities (Rumore et al. 2016).

With regards to climate change, role-play simulations can facilitate learning and engagement with the issue in various ways, including collectively considering local climate risks, acknowledging differing interests of various stakeholders in climate adaptation at the local level and experimenting with new policy tools and approaches without real-world political, social, or financial consequences (de Suarez et al. 2012; Rumore et al. 2016; Schenk 2015). The average citizen has limited incentive to and experience with incorporating the likely impacts of climate change in their daily lives (Leiserowitz et al. 2013). By introducing hypothetical municipal climate adaptation challenges to participants and pushing them to respond with and to other stakeholders, role-play simulations can foster an appreciation of planning and policymaking processes in which local adaptation decisions are made alongside competing policy priorities and limited resources (Schenk 2015).

Playing games is increasingly linked to several distinct learning outcomes. Playing games can result in four kinds of learning outcomes: cognitive learning outcomes, consisting of knowledge and cognitive skills; motor skills; affective learning outcomes; and communicative learning outcomes (Wouters et al. 2009). Despite the optimism about the potential of games for learning, some have noted that there is insufficient empirical evidence to support most of these claims (Bellotti et al. 2013; Gosen and Washbush 2004; Wu and Lee 2015), calling for more rigorous studies examining the effectiveness of games in learning about climate change and fostering readiness to climate risks, particularly those that compare game-based participation to other more commonly used engagement tools and methods (de Suarez et al. 2012).

From 2013 to 2014, the New England Climate Adaptation Project (NECAP) tested the effectiveness of role-play simulations as a civic education and engagement tool for climate adaptation with over five hundred participants in four towns in coastal New England. Participation in the NECAP role-play simulations was found to be associated with increases in concern for local climate change risks, sense of need for local adaptation action, and confidence in the prospects of effective local adaptation action (Rumore et al. 2016; Susskind et al. 2015). NECAP developed and tested four role-play simulations, each tailored to the specific climate risks and policymaking environments of each town.

The *Frames and Games* project expanded on the findings of NECAP by exploring the potential of using face-to-face multi-stakeholder role-play simulations highlighting the likely health outcomes from climate change to mobilize publics to respond to climate risks at the local level. To date, a role-play simulation exercise that engages participants to learn about the local health impacts of climate change has not yet been designed and implemented. Similarly, there is a lack of empirical studies measuring the efficacy of this approach to bolster public awareness of shared climate risks and enhance support for local climate adaptation action.

A Role-Play Simulation on the Health Impacts of Climate Change

The face-to-face role-play simulation[1] used in the *Frames and Games* research project is a game set in the fictional city of Mapleton with six different assigned roles: City Manager, the Director of Public Health, the Director of Public Housing, President of the local Chamber of Commerce, Executive Director of a local environmental advocacy organization, and a local university planner. With the help of a neutral facilitator, participants debate and recommend policy decisions to local officials in Mapleton to help prepare the city for the likely health impacts of climate change. Each participant assumed one of six roles in the game. Every participant was given a set of common general instructions

1 The role-play simulation was designed according to best practices in role-play simulation design, building on over a decade's experience at the MIT Science Impact Collaborative and the Consensus Building Institute. The role-play simulation was tailored to the specifics of Cambridge, integrating the likely local health impacts of climate change into the role-play simulation, based on the results of the Climate Change Vulnerability Assessment undertaken by the City of Cambridge. The role-play simulation materials are available through the Harvard-MIT-Tufts Program on Negotiation at https://www.pon.harvard.edu/shop/how-to-handle-the-public-health-impacts-of-climate-change/.

and role-specific confidential instructions that outlined their role's interests and positions within a climate adaptation planning and policymaking process.

As the game featured Cambridge's heat and precipitation-driven flooding data from the city's actual climate projections, the role-play simulation induced participants to learn about local climate risks in their own municipality. Participants also learned about the adverse health outcomes likely to occur in the city due to the projected increases in frequency and duration of extreme heat events and increases in the frequency and intensity of extreme precipitation events. Finally, the role-play simulation presented a suite of policy approaches available to local governments to alleviate the health risks of climate change as a starting point for the role-play simulation groups' deliberation of potential strategies the hypothetical municipality could undertake. Participants were explicitly encouraged to engage in policy experimentation through modifying the existing policy approaches and creating hybrid strategies. The role-play simulation's instructions presented each of the five policy approaches with a description of the activities associated with the approach and its predicted public health outcomes.

Playing the role-play simulation was associated with *more* (1) perceived knowledge of climate change, (2) concern for local climate impacts, and (3) confidence in the city's effectiveness in local climate adaptation action. While the role-play simulation attracted the smallest number of politically conservative people (versus moderate or liberal), the conservatives that attended showed the *largest increases* in (1) perceived knowledge of climate change, (2) concern for local climate impacts, and (3) confidence in the city's prospects for effective local climate adaptation. In other words, the face-to-face role-play simulation effected transformative counter-attitudinal learning.

In addition, qualitative data collected during and after the role-play simulation point to the potential of role-play simulations to facilitate social learning in terms of (1) municipal planning and policymaking process-based knowledge, (2) perspective-taking of different stakeholders in climate adaptation at the local level, and (3) strengthening of the place identity and attachment of participants. The following sections present more details on the research process and the results of the face-to-face role-play simulation.

Research Process

From November 2015 to May 2016, 118 participants played the face-to-face role-play simulation in six workshops. Before participating in the role-play simulation workshops, people completed a before-survey with questions measuring their perceived knowledge of climate risks, their concern for climate impacts, and their policy priorities for local climate adaptation, among other things. The before-survey questions are in Appendix C. After playing the role-play simulation, the participants completed an after-survey to gauge the efficacy of the role-play simulation in effecting change in cognition, affect, and behavior around local climate action. The survey also posed questions about demographics and other characteristics correlated with environmental and climate change–related cognition and behavior in past research, with a focus on parsing how participants of differing political viewpoints (conservative, moderate, liberal) showed differences in learning from the role-play simulation experience. The after-survey questions are in Appendix D.

Both surveys posed the same questions on cognition, affect, and behavioral intentions[2] around climate change at the local level to measure shifts from participating in the face-to-face role-play simulation highlighting the likely health impacts of climate change in their municipality in the following variables:

a) perceived knowledge of climate change[3] [cognition]
b) perceived seriousness of threats from climate change[4] [affect]
c) concern for climate impacts in their municipality[5] [affect]
d) confidence in effective local climate adaptation action[6] [affect]
e) support for city incorporating climate risks into everyday planning[7] [behavior]
f) willingness to pay higher taxes for their city to reduce climate risks[8] [behavior]
g) climate risk the local government should prioritize most in its climate adaptation policies [behavior]
h) preferred policy approach to manage the health impacts of climate change [behavior]

Expected results

Compared to before participating in the face-to-face role-play simulation, participants were expected to have, on average, after playing the role-play simulation:

a) a higher level of perceived knowledge of climate change [cognition]
b) a higher level of perceived seriousness of threats from climate change [affect]
c) a higher level of concern for climate impacts in their municipality [affect]
d) a higher level of confidence in effective local climate adaptation action [affect]
e) a higher level of support for city incorporating climate risks into everyday planning [behavior]

---

2 In social psychology, attitude is defined as "an individual's disposition to respond favorably or unfavorably to an object, person, institution, or event, or to any other discriminable aspect of the individual's world" (Ajzen 1989, 241). Rosenberg and Hovland (1960) posit in their tripartite model of attitude structure that one's attitude can be inferred by measuring three components of cognition, affect, and behavior. This tripartite model of attitude structure is the most prevalent conceptual model of attitude in social psychology research (see Ajzen 1989; Allport 1964; Hilgard 1980; McGuire 1969 for a review). See Bagozzi et al. (1979), Breckler (1984), and Eagly and Chaiken (2007) for a discussion regarding the construct validity of the tripartite classification of attitudes.

3 Validated and tested questions from Lee et al. (2015).

4 Validated and tested questions from Lee et al. (2015).

5 Validated and tested questions from Rumore et al. (2016).

6 Validated and tested questions from Rumore et al. (2016).

7 Validated and tested questions from Rumore et al. (2016).

8 Question adapted from the General Social Survey (GSS) on willingness to pay higher taxes for environmental protection

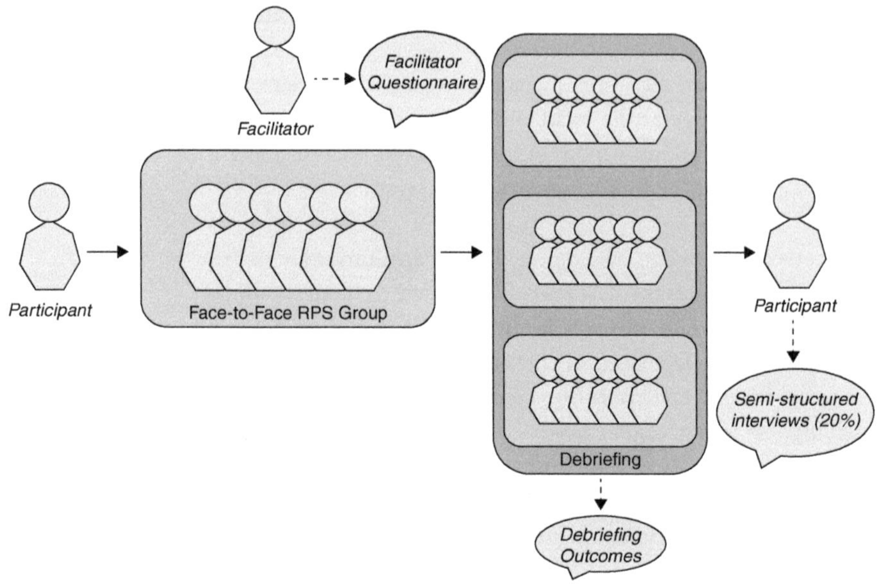

Figure 5.1 Overview of face-to-face role-play simulation and data collection

f) a higher level of willingness to pay higher local taxes for their city to reduce climate risks [behavior]

g) a higher proportion of the respondents picking "increased risks of disease, hospitalization, and death" as the climate risk the local government should prioritize in its climate adaptation policies [behavior]

h) a higher proportion of the respondents picking "emergency preparedness" as the policy approach the local government should primarily use to manage the health impacts of climate change [behavior]

Additional data collected during and after the role-play simulation were used to triangulate the before- and after-survey results (Creswell and Clark 2007) and provide a richer narrative with qualitative evidence on how the role-play simulation impacted participants' views of climate risks and local climate adaptation planning and policymaking efforts.

Figure 5.1 presents an overview of the face-to-face role-play simulation, along with the three additional data sources from the role-play simulation workshops analyzed in this research project, together with the before- and after-surveys.

First, each role-play simulation group (consisting of six participants) had a trained facilitator aiding the group in collectively constructing a policy recommendation. Each facilitator filled out a questionnaire with their observations of their group's outcomes, which process- or substance-related arguments carried weight in their group, and how different stakeholders worked to craft the final policy bundle of choice within the role-play simulation. The facilitator questionnaire is in Appendix E.

Second, the game participants at the workshop gathered as an overall group after the role-play experience to debrief the game results and step back into their real-life roles as citizens of Cambridge to discuss how the game principles apply to their municipality. Extensive notes were taken during the debriefing for subsequent qualitative analysis. During the debriefings, participants reflected on what happened at their group in terms of process and substance, compared their results to that of other groups within the entire workshop group, and related the game to their own realities in in terms of local climate risks and policy options for climate adaptation action in Cambridge. The debriefing script is in Appendix F.

Third, semi-structured interviews were conducted[9] with about 20 percent of the role-play simulation participants to not only measure learning outcomes over time but also examine if and how the face-to-face role-play simulation modified their perceptions of climate risks and local climate adaptation policy priorities. The semi-structured interview questions are in Appendix G.

Results

Compared with Cambridge's citywide statistics, the role-play simulation participant sample skewed to older, more politically liberal, more educated, and relatively higher-income people.[10] Appendix H presents a detailed comparison of the role-play simulation participant sample and the city of Cambridge's demographics.

Role-play simulation outcomes

A total of 20 groups participated in the face-to-face role-play simulation. Out of the 20 groups, 12 groups reached a unanimous agreement on a final policy recommendation to city officials in Mapleton within the allotted game sessions, while five groups had five out of six participants agree with the final recommendations put forth by their group, and three out of the 20 groups did not reach an agreement. In terms of the policy approaches selected to manage the health impacts of climate change, emergency preparedness was the most popular (15; out of 20 groups total), and prevention was a close second (12; out of 20 groups total).

---

9 The interviews typically took about 30 minutes to an hour each. The interviews were recorded, transcribed, and coded to identify patterns within interview responses and extract quotes exemplifying key themes.

10 Recruiting difficulties, along with the critiques that public engagement workshop participants are not demographically representative of the general population, have highlighted challenges with scaling up face-to-face role-play simulations (RPS) for larger audiences. Additional time and resources for recruitment would enable more robust sampling processes in future research.

Themes from the debriefings: Connecting game results to real-life situations

Several themes emerged during the debrief sessions of the role-play simulation workshops, in which participants related the game dynamics to their real-life situations as citizens in Cambridge.

1. **Approximately half of the participants indicated that they had never thought about the public health impacts of climate change prior to the role-play simulations.**

   For those more skeptical or less willing to act on the findings of environmental science, alleviating the health impacts of climate change offered a tangible benefit of taking action, facilitating greater consensus among participants. One participant commented that focusing on public health helped to connect climate research and activism by blending science, policy, and social justice. Furthermore, because health is something all people can relate to and identify with, a focus on public health adds value to public engagement on climate policy by making the impacts of climate change seem more tangible.
2. **Participants argued that a shift in public opinion of climate change is necessary to facilitate climate adaptation.**

   In three of the six workshops held, at least one participant noted the similarities in shifting the public perception of climate risks and smoking in the twentieth century. It takes substantial effort and time to alter public perceptions of health risks and encourage people to adjust their personal and political choices. In this sense, participants thought using the analogy of smoking to highlight the health risks of climate change could help illustrate the dangers of inaction in climate adaptation.
3. **Public health is an effective framing mechanism to motivate climate action.**

   Participants discussed that public health is an effective frame for communicating the severity of climate change and the urgency to mitigate and prepare for it. They noted the positive ways in which the frame communicates that climate change is a local issue and how the frame connects it to quality-of-life issues, which are easier for the general public to understand. This frame can help with outreach and connecting to more members of the population. In addition, the public health frame of climate change emphasizes the negative health outcomes occurring *now*, heightening a sense of temporal proximity.
4. **A public health focus of climate change highlights the disproportionate impacts on low-income and vulnerable populations.**

   Participants mentioned that focusing on public health was a relevant way of thinking about how climate change will affect these populations and how public policies and investments are critical to protecting the most vulnerable community members from the health risks of climate change. Some participants expressed that they have the social and financial capital to reduce risks for themselves but are concerned about their neighbors who do not. Thus, they argued that the local government should help the most vulnerable prepare for climate change, and this shared responsibility should be reflected in a common financial burden to ensure collective well-being.

5. **Participants recognized that climate change is causing both short- and long-term impacts on health and expressed a near universal acceptance that local climate adaptation planning and policymaking processes should seek to mitigate the anticipated short-term public health impacts of climate change while planning for long-term community needs in the context of shifting environmental conditions.**

   One participant noted that the focus on public health highlighted the need to think about short-term responses to the health risks today and not just the long-term mitigation of climate change likely to arise in the future. All groups included some focus on emergency preparedness to address the short-term impacts on health, and most included additional strategies to mitigate longer-term impacts.

6. **Participants believed that climate adaptation would be most effective if the city works with existing community organizations, such as institutions of faith, community-based organizations, and tenant associations to strengthen climate resilience and improve public health outcomes.**

   In addition, climate change poses an opportunity for enhancing social resilience to climate change. Many of the creative strategies that emerged from the role-play simulations included collaborative initiatives with local businesses and universities to spur innovation in climate adaptation. Furthermore, participants also noted that as health is already a focus across different government functions and scales, a focus on the health impacts of climate change can provide a substantive focal point to "mainstream" climate change with across different government agencies.

7. **Participants argued that local governments and citizens need to be well-informed of the local climate risks and relevant health impacts to make appropriate decisions.**

   A focus on public health also drives an interest in city-supported means for helping the public protect themselves. Many participants noted that public information is a critical component of emergency preparedness.

8. **Many participants discussed the importance of co-benefits, such as public space improvements with health benefits, in identifying strategies to recommend.**

   Strategies with co-benefits address more than one issue and can be viewed as win-wins. Participants stressed prioritizing the numerous opportunities to adapt while increasing social capital and beauty. Other strategies with co-benefits suggested were green infrastructure projects and partnering with universities to run pilot programs that could be replicated elsewhere in the city. Participants agreed that Cambridge should strive to find solutions that build on community assets, improve the urban landscape, and make the city a more enjoyable and desirable place to live.

Group decision-making success factors

According to the facilitators that guided the groups through the role-play simulation exercise, several substance and process factors stood out in terms of their impact on their groups' outcomes.

With process, many facilitators noted that their groups mostly consisted of participants invested in the role-play simulation and in their group's success in the role-play simulation. One facilitator remarked that he had worked with a "congenial group overall," and his group shared an "early agreement that *something* had to be done." Another facilitator stated that her group had been "generally cooperative and interested in reaching consensus," and "coalitions formed early on in the group for each of the two [policy] priorities that prevailed, which helped focus the discussion."

In addition to the facilitators, some groups had participants exhibiting political leadership or "people help[ing] shepherd others to [the] outcome." One facilitator described the "shepherding" as the following: "A couple of proactive participants would note out loud when it seemed like there was relative consensus (or minimal opposition) to a particular strategy as we moved down the list, making the ultimate process of formulating agreement relatively smooth and easy because they had already noted a number of common interests and positions."

All the groups that had successfully reached an agreement modified the proposed policy approaches and creatively adapted the strategies to meet the group's interests, or, in one facilitator's words, engaged in the "hybridization and specialization of strategies." Another facilitator noted that the issue of financial feasibility carried significant weight at his table in reaching the final outcome, remarking that his table "only reached a solid agreement after mentioning the evolving or adapting nature of funding," and emphasized that his group's policy recommendations were very much reliant on funding outcomes.

Shifts in climate knowledge, affect, and behavior after playing the game

A series of paired t-tests[11] were run[12] on the sample of 112[13] role-play simulation participants to determine if there were statistically significant differences in knowledge, affect, and behavior around local climate adaptation before and after participating in the role-play simulation.

Table 5.1 sums up the main effects of the role-play simulation on key dependent variables. In line with expectations, role-play simulation participants showed statistically

---

11 Participants identified themselves on both surveys, allowing for matched-pairs analysis.

12 This study acknowledges the debate around using parametric tests on Likert scale data. While principled stances such as Jamieson (2004, 1217) forbid conducting parametric analyses on ordinal data as "intervals between values cannot be presumed equal," empirical studies of the robustness of tests of central tendency, including t-tests and ANOVA, find that parametric tests on differences in means for sample sizes greater than 5 show robustness with respect to ordinality and non-normality (Boneau 1960; Norman 2010; Pearson 1931). In addition, Wilcoxon signed-rank tests were also conducted with similar results, enhancing confidence in the parametric test outcomes.

13 Out of the total of 118 people who participated in the role-play simulations, after-surveys were collected for 112 participants; six surveys were not collected from participants that had either not completed the role-play simulation in its entirety or had chosen not to submit their after-surveys.

Table 5.1 Main Effects of RPS on Key Dependent Variables

Dependent variables (n = 112)	Role-Play Simulation Δ Post-Pre	Cohen's *d*
Perceived knowledge of climate change (Likert scale:1–3)	**0.073*** (0.031)	0.226
Perceived personal threats from climate change (Likert scale:1–5)	0.008 (0.051)	0.016
Concern for local climate risks (Likert scale:1–5)	**0.259**** (0.099)	0.247
Confidence in city's effectiveness in climate adaptation[a] (Likert scale:1–5)	**0.214*** (0.104)	0.275
Agreement with city incorporating climate risks into everyday planning (Likert scale:1–5)	**−0.128*** (0.058)	0.21
Willingness to pay higher taxes for climate adaptation (Likert scale:1–5)	−0.083 (0.076)	0.104

Note: Standard errors in parentheses. All mean comparisons significant at *p < 0.5 (bold face), **p < 0.1 (bold face), ***p < 0.01 (bold face). Unequal variances assumed (t-test). Cohen's *d* is a standardized measure of effect size; values between 0.3 and 0.6 are considered "moderate" effect sizes in behavioral sciences (Cohen 1988).
[a] n = 56.

significant increases in their (1) perceived knowledge of climate change, (2) concern for local climate risks, and (3) confidence in their municipality's ability to effectively plan for the impacts of climate change after playing the role-play simulation when compared to before participating in the role-play simulation. Playing the role-play simulation was not associated with statistically significantly higher levels of perceived personal threats from climate change, support for their city incorporating climate risks into everyday planning,[14] or willingness to pay higher taxes for their municipality to reduce climate risks than people exposed to the environmental frame of climate change.

Shifts in knowledge, affect, and behavior by political viewpoint

Table 5.2 shows the effects of the role-play simulation on key dependent variables by political viewpoint. Research has shown that political orientation is the biggest

[14] Interestingly, the participants also unexpectedly showed a statistically significant decrease in their average level of agreement with the following statement: "The city of Cambridge should incorporate projections of what the climate might be like in 50 years in everyday planning and infrastructure decisions." Does the statement signify a reliance on accurate and reliable climate data to emerge, and make people nervous about a delay in climate action? More research is called for to investigate why playing a role-play simulation that focuses on the health impacts of climate change within their community might result in feeling less strongly about local governments using climate projections in their planning processes.

Table 5.2 Main Effects of RPS on Key Dependent Variables by Political Viewpoint

Baseline Pre-RPS Means (Average Treatment Effect)	Average (n = 112)	Conservative	Moderate	Liberal
Perceived knowledge of climate change* (Likert scale: 1–3)	2.624 **(+0.073)**	2.2 **(+0.2)**	2.57 **(+0.143)**	2.64 **(+0.07)**
Perceived personal threats from climate change (Likert scale: 1–5)	4.446 (+0.009)	3.6 (+0.0)	4.14 (+0.143)	4.55 (+0.022)
Concern for local climate risks** (Likert scale: 1–5)	3.901 **(+0.259)**	3.2 **(+0.2)**	3.43 **(+0.143)**	4.0 **(+0.281)**
Confidence in city's effectiveness in climate adaptation*[a] (Likert scale: 1–5)	2.554 **(+0.214)**	2.0 **(+0.33)**	2.75 **(+0.25)**	2.61 **(+0.2)**
Agreement with city incorporating climate risks into everyday planning (Likert scale: 1–5)	4.8 (−0.128)	4.6 (−0.4)	4.86 (−0.43)	4.84 (−0.11)
Willingness to pay higher taxes for climate adaptation (Likert scale: 1–5)	3.862 (−0.083)	3.2 (−0.4)	3.0 (−0.143)	3.99 (−0.081)

Note: Dependent variables with asterisks have mean increases significant at $*p < 0.5$, $**p < 0.1$, $***p < 0.01$.
[a] n = 56.

determinant of belief in climate change (Hornsey et al. 2016) and that people are likely to selectively pick data most consistent with their political outlooks (Kahan et al. 2012; Kahan et al. 2013). Accordingly, identifying public engagement tools in climate planning and policymaking processes that lead to changes not only in moderates and liberals but also in conservatives is critical to alleviating the "echo chambers" of climate change (Williams et al. 2015) and furthering climate adaptation action (van der Linden et al. 2016).

In terms of baseline means before playing the role-play simulation, conservative participants had the lowest means for perceived knowledge of climate change (2.2 versus 2.57 and 2.64), concern for local climate risks (3.2 versus 3.43 and 4.0), and confidence in their municipality to effectively plan for and respond to climate change (2.0 versus 2.75 and 2.61). However, politically conservative participants also showed the largest increases in perceived knowledge of climate change (0.2 versus 0.143 and 0.07), concern for local climate risks (0.2 versus 0.143 and 0.281), and confidence in the prospects of effective local climate adaptation action (0.33 versus 0.25 and 0.2). Therefore, the role-play simulation was not only effective in increasing conservative participants' thinking and feeling around the need for climate adaptation at the local level, but it was also the *most* effective among conservatives, when compared to its effects on politically moderate and liberal participants.

This finding points to the potential of using a role-play simulation on local climate-related health risk management to transcend the growing political polarization on climate change (Dunlap et al. 2016), and its significance cannot be understated.

Table 5.3 Effects of RPS on Different Types of Learning by Political Viewpoint

Perceived Learning (1–5)	Average (n = 112)	Conservative	Moderate	Liberal
Health impacts of climate change	2.79	2.33	2.75	2.83
Views of other stakeholders in local climate adaptation	2.83	2.33	3.25	2.83
Reaching an agreement in local climate adaptation	2.8	2	3	2.83

Different types of learning in the role-play simulation

Table 5.3 depicts how game participants ranked the relative amounts of learning in different dimensions of climate adaptation from the role-play simulation. In the first column (Average) of Table 5.3, the average role-play participant said the role-play simulation helped them the most in learning about the viewpoints of different stakeholders in dealing with climate change in their municipality (2.83), then imagining how different stakeholders could reach an agreement on how to proceed with climate adaptation at the local level (2.8), and, finally, the health impacts of climate change. It is interesting to note that participants felt they learned the most about the views of other stakeholders in municipal planning and policy-making processes, and the least about the health impacts of climate change (2.79).

For the substantive learning about the health impacts of climate change, politically liberal participants said they learned the most, then the moderates, and finally the conservatives. For learning about the various views of stakeholders in local climate adaptation efforts, however, moderates said they learned the most and conservatives the least. Finally, for improving people's ability to imagine how stakeholders could reach an agreement on how to proceed with climate adaptation at the local level, there is a similar pattern, where politically moderate participants experienced the most learning, then liberals, and finally conservatives. These differences are visualized in Figure 5.2.

Interestingly, when asked directly how much they learned from the game, conservative game participants consistently stated on the after-surveys that they learned less than the participants of other political viewpoints. This only accentuates the relatively large shifts in conservatives' perceived knowledge of climate change, concern for local climate risks, and confidence in their local government's ability to effectively plan for the impacts of climate change after playing the role-play simulation, when compared to the shifts seen in moderate and liberal participants. These results suggest that conservative game participants may have discounted the amount of learning from the material presented in the role-play simulation, especially if it countered their own political views, but, nonetheless, experienced the largest increases in their urgency around local climate risks and commitment to future climate adaptation action.

Shifts in local climate adaptation policy priorities

Playing the role-play simulation not only enabled substantive and process-based learning and increased perceived knowledge of and concern about climate risks and confidence

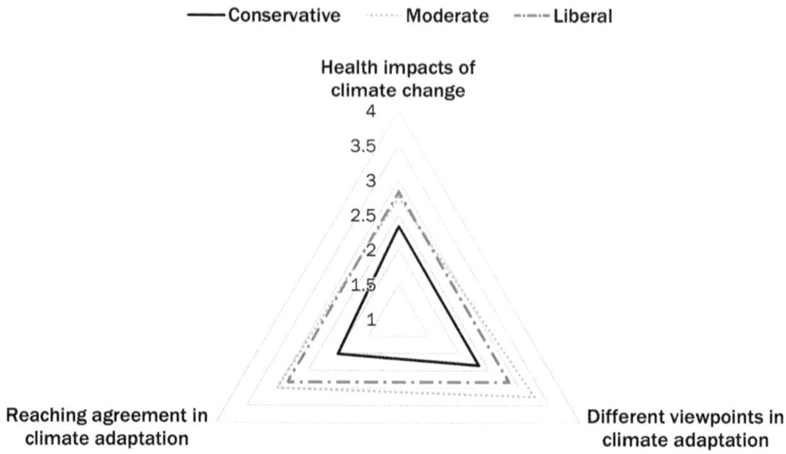

Figure 5.2 Effects of RPS on different types of learning by political viewpoint

in local climate action but also shifted participants' policy priorities on the actions their city should take to adapt to climate change.

Table 5.4 depicts the change in the proportions of respondents' prioritization of local climate risks to focus on in their city's climate adaptation processes, before and after the role-play simulation. The percentages represent the proportion of the group that selected each risk as their first priority in their city's climate risk–management programs. There is a statistically significant shift in the game participants' prioritizing the increased risks of disease and death, going from 14.0 percent to 41.9 percent, with a simultaneous decrease in prioritizing infrastructure damage from 60.5 percent to 34.9 percent, after the role-play simulation.[15] By emphasizing the health implications of climate change at the local level, the role-play simulation shifts people's attention to policy approaches that directly aim to reduce adverse health outcomes from and increase human resilience to climate change.

Shifts in preferred policy approaches to manage local climate risks

Furthermore, the role-play simulation experience also changed people's preferred policy options to reduce climate-related risks in their municipality. Table 5.5 depicts the shift in the proportions of respondents' prioritization of policy approaches to manage local

[15] All other marginal proportions are similar before and after the intervention. Results of a McNemar's test show that the marginal proportions are statistically significantly different from each other (Fisher's exact = 0.002). The McNemar's test is a statistical test used on contingency tables with matched pairs of subjects, to determine whether the row and column marginal frequencies are equal. If the Fisher's exact score low enough ($p < 0.05$), this provides sufficient evidence to reject the null hypothesis and state that the marginal proportions are statistically significantly different from each other.

Table 5.4 Prioritization of Local Climate Risks in Climate Adaptation, Before and After RPS

	Before	After
Individual property damage	2.3%	2.3%
Local business disruption	0.0%	0.0%
Increased risks of disease, hospitalization, and death	14.0%	41.9%
Disproportionate impacts to vulnerable populations	16.3%	16.3%
Water, energy, and transportation infrastructure damage	60.5%	34.9%
Watershed, forest, and ecosystem degradation	7.0%	4.7%
Total	100%	

Table 5.5 Prioritization of Policy Approaches to Manage Climate Risks, Before and After RPS

	Before	After
Information provision	12.9%	12.9%
Emergency preparedness	9.7%	30.7%
Resource allocation	8.1%	4.8%
Rules and regulations	9.7%	17.7%
Prevention	59.7%	33.9%
Total	100.0%	

climate risks, before and after the role-play simulation. Similarly, the percentages represent the proportion of the group that selected each policy approach as their first priority in their city's climate policy. There is a sizable increase in participants' preferences for emergency preparedness as the preferred policy approach (30.7 percent versus 9.7 percent), with a smaller increase in rules and regulations (17.7 percent versus 9.7 percent). There is also a significant decrease in people ranking prevention as best policy approach, going from 59.7 percent to 33.9 percent.[16]

This increased policy attention to emergency preparedness after learning about the managing the health impacts of climate change reflects an emphasis on short-term responses to current climate-related health risks. Preparedness consists of devising plans or procedures to save lives and to minimize damage in adverse events, and good emergency preparedness ensures that when a disaster occurs, municipalities will be able to respond in optimal ways. The simultaneous decrease in the selection of prevention also highlights the critical need to balance trade-offs between short-term response capacities to minimize deaths and injuries and more upstream climate mitigation activities to reduce future climate-related health risks.

16 All other marginal proportions are similar before and after the intervention. Results of a McNemar's test show that the marginal proportions are statistically significantly different from each other (Fisher's exact = 0.003).

Inside the role-play simulation experience

A majority of the role-play simulation workshop participants interviewed[17] enjoyed the role-play simulation, saying it was informative and entertaining. Furthermore, some articulated in detail how the role-play simulation helped them learn about climate adaptation planning and policymaking processes at the local level.

Many participants remarked that through the role-play simulation experience, they gained process-based knowledge of how planning and policymaking works in a city like their own. One participant described the dynamics within her role-play group and how she learned to maneuver multiparty stakeholder dynamics in the public realm: "One person took their role very strongly, and was not going to be moved. I was struck by how effective that person was in blocking further progress." She then explained that her group persevered in trying to "find interesting strategies to attempt to join interests, rather than have something that was solely divisive." As a result, she "got a sense that even with intractability, you can work to find shared concerns." Another participant noted that he "found the discussion of figuring out how we were going to compromise" and "the give and take and forming alliances around issues to be very interesting." Finally, one participant remarked that "what playing the game did for me was to help me understand again how important activism is, because the wheels of bureaucracy don't turn quickly," and described how she reflected on her role within these processes in her city, noting that "the idea of doing these role-play processes is very powerful [...] what kind of an impact can we have?"

In addition, interviewees stated that the role-play simulation was a valuable perspective-taking exercise. One participant elaborated on the process, stating, "Being there, being a participant made you set aside your own perspective and think about it from the character you were playing." Another participant said, "I think the low-income housing representative had strong feelings about things I wouldn't have thought were important before." Highlighting the power of playing a role different from your everyday job, an environmental advocate said that while playing an advocacy role for local businesses, "articulating the positions of the Chamber of Commerce forced me to learn about relevant interests."

Finally, participants remarked on how the role-play simulation helped them learn from and about fellow citizens of their municipality. Many were surprised to see a large number of people interested and engaged in the issues presented in the role-play simulation. One interviewee reflected,

> It highlighted there are a lot of interested stakeholders who want to be engaged, which is positive. Normally it is hard to find a lot of people engaged in science, but it seems like there are a lot of people that are interested in climate risks here, and that's something the city should take note of as they craft their preparedness plan, that they have a lot of stakeholder support.

---

[17] n = 21.

There were several participants noting the role-play simulation exercise strengthened their pride for their city, such as one remarking, "I have deep affection that this [role-play simulation workshop] is happening. I'm proud to be from here!" The role-play simulation can strengthen a city's collective efficacy in planning for climate change and reinforce people's place identity and attachment for future climate adaptation planning and policymaking efforts to build upon.

Certain elements of the role-play simulation elicited critical responses from the interviewees. As the role-play simulation is a simplification, some participants pointed out that certain elements felt less than real, especially in presenting goals of collaboration or a group gathering to reach an agreement. One participant explained, "In the context of any type of collaboration, it can only happen if parties are working in good faith. It's simply not possible to work with climate change deniers. That's the nature of the problem." Another participant said that she felt the role-play simulation was unrealistic because the goals were already set, as "it's really easy to work with people as long as the goal is set." These comments point to the importance of managing the trade-off between simplifying the role-play simulation to structure the experience within a limited amount of time and enhancing the game's fidelity to real-world circumstances.

In addition, one participant mentioned that she noticed a certain demographic to be more likely to attend the role-play simulation workshops, saying, "I don't think the people in the room represented a cross-section of 'Mapleton.' [...] I think everyone in the room was a fairly privileged, articulate, educated person. There was some racial diversity, but I don't think it is representative of our city." People with an existing interest in a topic are more likely to attend a face-to-face event to engage with other individuals who are similarly able and willing to commit the necessary time and resources, suggesting that the role-play simulation itself might be limited in its potential to address the root causes of the "usual suspects" dilemma in municipal planning and policymaking processes.

Conclusion

Playing the role-play simulation was associated with increases in perceived knowledge of climate change, concern for local climate impacts, and confidence in the prospects of effective local climate adaptation action. While the role-play simulation attracted the smaller number of politically conservative people (versus moderate or liberal), the conservatives that did participate showed the largest increases in perceived knowledge of climate change, concern for local climate impacts, and confidence in the city's effectiveness in local climate adaptation. In other words, the face-to-face role-play simulation effected transformative counter-attitudinal learning.

In addition, qualitative data collected during and after the role-play simulation point to the potential of role-play simulations to facilitate learning how municipal planning and policymaking processes work, perspective-taking of different stakeholders in climate adaptation at the local level, and strengthening the place identity and attachment of participants.

Chapter 6

DIGITAL GAMES

The Promise of Digital Games

Local governments need to devote considerable money, time, and attention to civic ties for traditional public engagement practices (Lowndes et al. 2001). However, conventional public meetings, or any other institutionalization of public engagement, are limited in the extent to which they can serve as a learning venue where the public can deepen their understanding of a policy issue (Evans-Cowley and Hollander 2010). Mostly, people attend these events to express their views and hear the views of others. As a result, much attention is now being given to engaging the public online, as digital engagement requires less money and time when compared to traditional face-to-face events (Best and Krueger 2005). In addition, promises of online forms of public engagement's ability to "elevat[e] the public discourse in an unprecedented manner while providing an interactive, networked environment for decision-making" abound (Evans-Cowley and Hollander 2010, 397).

In parallel, the education field has turned to various forms of engaging and educating people online. Digital games are hugely popular. In the United States, 63 percent of households has at least one person who plays video games regularly, or 3 hours or more per week (ESA 2017). Digital learning games are defined as "an entertainment medium designed to bring about cognitive changes in its players" (Erhel and Jamet 2013, 156), characterized by having both elements of serious learning and interactive entertainment (Erhel and Jamet 2013; Prensky 2003). The presence of rules and constraints, dynamic responses to players' actions, appropriate challenges inducing perceived self-efficacy, and gradual increases in difficulty have been found to contribute to digital game-based learning (Mayer and Johnson 2010).

Whether digital games can and do actually facilitate deep learning is under debate (Graesser et al. 2009). According to Graesser et al., digital games have innate constraints that "make it extremely difficult to integrate deep content, strategies, and skills" (2009, 12). Scholars have called for more empirical analyses that examine the efficacy of digital games in catalyzing motivation and learning outcomes (Connolly et al. 2012; Gee 2003; Prensky 2003; Shaffer 2006). While digital games' efficacy in motivating players is less contested, how much digital games contribute to players' learning, especially in comparison with more conventional teaching methods, has yet to be resolved (Erhel and Jamet 2013). Research shows that the learning effects of digital games are modulated by various factors, including individual player characteristics, such as prior knowledge, motivation levels, or learning styles (Tsai et al. 2012; Vogel et al. 2006), learning environment features (Erhel and Jamet 2013; Ke 2009; Vogel et al. 2006), and subject matter (Hays 2005; Huang 2011; Ke 2009; Randel et al. 1992).

Digital Games and Climate Change

There are now many digital learning games that focus on climate change. A review conducted in 2015 found a majority of these games to be mini-games or simple simulations, geared toward younger audiences, such as NASA's *Climate Kids* (http://climatekids.nasa.gov). Most of these are also said to focus on climate mitigation practices (i.e., reducing greenhouse gas emissions) that are implementable in daily life, such as recycling or using public transit (Wu and Lee 2015).

On the other hand, there are a few serious climate change games geared toward adults that are quite complex. These center on understanding the long-term impacts of climate change and emphasize the geophysical and sociopolitical dimensions of undertaking climate action. Wu and Lee (2015) cite *Clim'way* as an example, which is akin to a Sim City for climate change. Players make decisions regarding infrastructure in their cities and see the impacts of their actions over the course of 50 years. In Wu and Lee's review, all the digital learning games they discussed were designed to facilitate learning about climate change as a phenomenon or climate mitigation practices in daily life. Not one digital learning game was aimed at increasing knowledge of climate risks in one's own city or climate adaptation planning and policymaking processes at the local level (Wu and Lee 2015).

Lee et al. (2013) summarize research on the factors contributing to effective digital learning games on climate change. In particular, they emphasize the importance of presenting actionable knowledge to induce behavioral changes. They argue that increasing understanding of the broad conceptions of climate change does not tend to lead to differences in game players' lives and that digital learning games should tangibly connect the elements of climate change with realistic steps players can take to change their lifestyles (Lee et al. 2013). In addition, digital games can effect behavior change by fostering empathy in game players. Kim et al. (2010) found that game participants were more willing to reduce home electricity usage after seeing the impacts of their behavior on a virtual coral reef's health.

Based on previous findings regarding effective digital game-based learning in climate change, the potential for digital games to enhance public engagement, and the apparent lack of digital games focused on climate adaptation, this study designed an interactive place-based game with a focus on fostering empathy and presenting actionable steps for players. The digital game also highlights the health impacts of climate change and the potential for public health-oriented climate adaptation efforts at the local level.

A Digital Game on the Health Impacts of Climate Change

The digital game used in the *Frames and Games* research project was developed with three aims: (1) reach a broader public than is possible with conventional public engagement

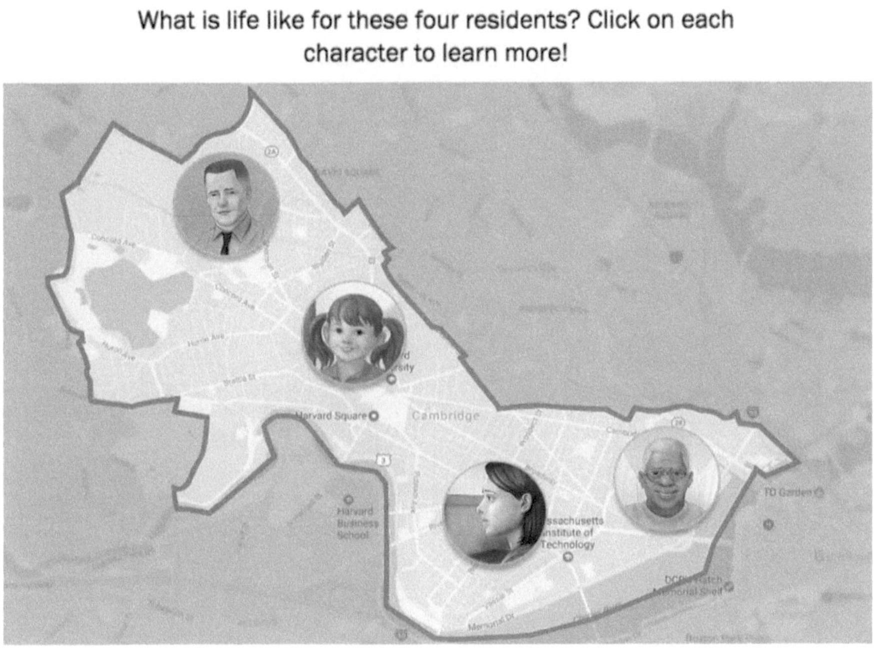

Figure 6.1 Home screen of digital game (from left to right: Jake, Cora, Sofia, Martin)

workshops or face-to-face role-play simulation workshops; (2) utilize in-game mechanics to assess changes in the cognition, affect, and behavior of game players; and (3) collect and process participant learning data with greater speed and quality.

In the digital game, participants assume the roles of hypothetical citizens in Cambridge who are more at risk to the health impacts of climate change because of where they live, their age, income level, or occupation. The game focuses on the increased health risks posed by rising temperatures, reduced air quality, increasing extreme weather, expanding water-related illness, and decreased food safety. The US Environmental Protection Agency (EPA) highlighted these risks in a series of fact sheets published to enhance public awareness and understanding of the health impacts of climate change (EPA 2016).

Figure 6.1 shows the home screen of the digital game with the four characters. The participants can choose to delve into the characters in any order of their choosing. Each character represents a unique subpopulation vulnerable to a specific climate-related health risk.

Aligning with best practices for digital game-based learning in climate change, this game focuses on providing personal accounts of the health impacts of climate change, along with actionable steps players can take in their everyday lives to reduce the relevant health risks or actions players can take in a caretaker role to someone who faces these increased health risks.

Each character is presented with three phases, each with distinct learning goals in mind. The first phase introduces the character and their existing health conditions, the second explains how climate change could impact the character through increased

Figure 6.2 Phase 1 of Martin

health risks, and the third and final phase features various readiness "tools" or actions, with corresponding explanations, that represent ways of alleviating some of the risks for the chosen character. After playing each character, the game presents one true/false question to determine whether the game player learned about climate risks and how they might affect the health of the character the player just delved into.

Figures 6.2, 6.3, and 6.4 show selected screens in each of the three phases for Martin, a retired architect with a history of heart problems, facing increased health risks from more frequent and intense periods of extreme heat. The other Cambridge citizens featured in the game are Cora, a young child with asthma; Jake, an amputee veteran; and Sofia, a recent immigrant. Each of these characters faces increased health risks caused by climate change.

Playing the digital game was associated with *more* perceived knowledge of climate change, concern for local climate impacts, and willingness to pay higher taxes for local climate adaptation action. The following sections go into more depth on the research process and the results of the digital game.

Research Process

From March 2017 to May 2017, during 57 days of game play, 229 players[1] completed more than 10,400 game actions. The digital game started by asking participants to fill

1 These participants were primarily recruited online, either directly through social media platforms or through email advertisements from local partners and partner organizations and various neighborhood organizations.

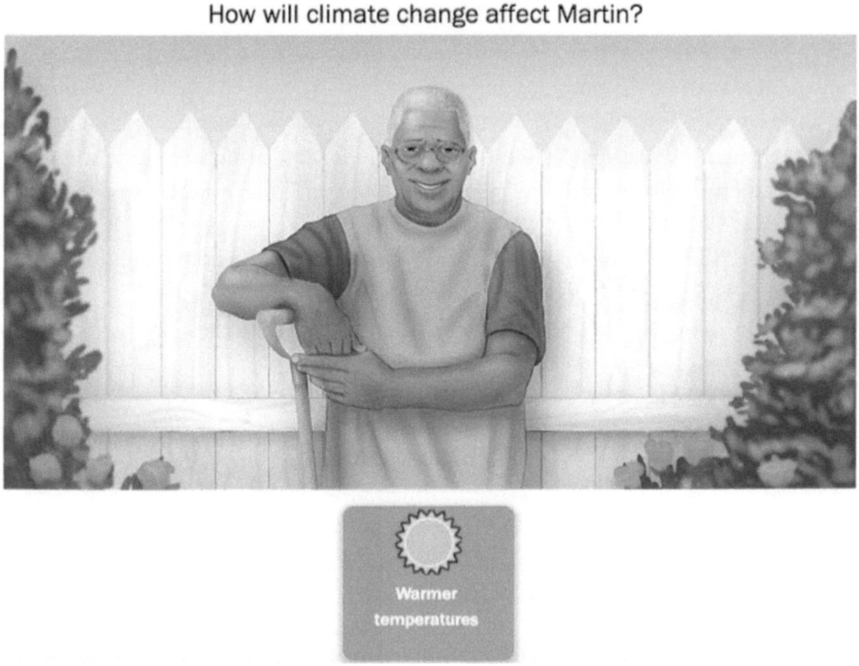

Figure 6.3 Phase 2 of Martin

Figure 6.4 Phase 3 of Martin

out a before-survey of questions documenting their perceived knowledge of climate risks, their concern about climate impacts, and their policy priorities regarding local climate adaptation, among other things. The before-survey questions are in Appendix I. After the completion of the digital game experience, the game platform posed an after-survey gauging the extent to which the game shifted levels of local climate change cognition, affect, and behavior. The after-survey also probed demographic variables and other correlates with environmental and climate change-related cognition and behavior. Finally, the last question asked for advice from game players regarding climate adaptation within their municipality: "What specific actions would you recommend the city of Cambridge take to manage the risks of climate change?" The after-survey questions are in Appendix J.

Both surveys posed the same questions on cognition, affect, and behavioral intentions[2] around climate change at the local level to measure shifts from playing the digital game, highlighting the likely health impacts of climate change in their municipality in the following variables:

a) perceived knowledge of climate change[3] [cognition]
b) perceived seriousness of threats from climate change[4] [affect]
c) concern for climate impacts in their municipality[5] [affect]
d) willingness to pay higher taxes for their city to reduce climate risks[6] [behavior]
e) climate risk the local government should prioritize most in its climate adaptation policies [behavior]

Expected results

Compared to before playing the digital game, participants were expected to have, on average, after playing the digital game:

a) a higher level of perceived knowledge of climate change [cognition]
b) a higher level of perceived seriousness of threats from climate change [affect]
c) a higher level of concern for climate impacts in their municipality [affect]
d) a higher level of willingness to pay higher local taxes for their city to reduce climate risks [behavior]
e) a higher proportion of the respondents picking "increased risks of disease, hospitalization, and death" as the climate risk the local government should prioritize in its climate adaptation policies [behavior]

---

2 In social psychology, attitude is defined as "an individual's disposition to respond favorably or unfavorably to an object, person, institution, or event, or to any other discriminable aspect of the individual's world" (Ajzen 1989, 241). Rosenberg and Hovland (1960) posit in their tripartite model of attitude structure that one's attitude can be inferred by measuring three components of cognition, affect, and behavior. This tripartite model of attitude structure is the most prevalent conceptual model of attitude in social psychology research (see Ajzen 1989; Allport 1964; Hilgard 1980; McGuire 1969 for a review). See Bagozzi et al. (1979), Breckler (1984), and Eagly and Chaiken (2007) for a discussion regarding the construct validity of the tripartite classification of attitudes.

3 Validated and tested questions from Lee et al. (2015).

4 Validated and tested questions from Lee et al. (2015).

5 Validated and tested questions from Rumore et al. (2016).

6 Question adapted from the General Social Survey (GSS) on willingness to pay higher taxes for environmental protection.

In addition to comparing the before- and after-survey results to determine if there were statistically significant differences in knowledge and concern about local climate risks and support for climate adaptation after playing the digital game focusing on the health impacts of climate change, (1) game mechanics, (2) in-game qualitative comments, and (3) feedback on the game were used to assess the learning outcomes of playing the digital game. A particular emphasis was placed on how participants of differing political viewpoints (conservative, moderate, liberal) showed differences in learning through the digital game experience. Finally, semi-structured interviews were carried out with digital game participants to examine if and how the digital game modified their perceptions of climate risks and local climate adaptation priorities. The semi-structured interview questions are in Appendix K.

Results

A unique ID was randomly generated for each IP address logged onto the game. Out of the total of 229 people who logged onto the game interface, 34 (14.9 percent) answered one or more of the before-survey questions but played 0 of the 4 characters, 46 (20.1 percent) played 1 of the 4 characters, 15 (6.6 percent) played 2 out of 4 characters, 5 (2.2 percent) played 3 out of 4 characters, and 129 (56.3 percent) played all 4 characters to complete the game. Out of the 129 people who completed the game, 105 players completed the after-survey, providing demographic information and comparison points to their before-survey responses.

Compared with Cambridge's citywide statistics, the digital game participant sample skewed to older, more politically liberal, and more educated people. Appendix L presents a detailed comparison of the digital game participant sample and the city of Cambridge's demographics.

Shifts in climate knowledge, affect, and behavior after playing the game

A series of paired t-tests[7] were conducted on the sample of up to 129[8] digital game participants that played all four characters to complete the game, and also completed both the before-survey and the after-survey, to determine if there were statistically significant

---

[7] This study acknowledges the debate around using parametric tests on Likert scale data. While principled stances such as Jamieson (2004, 1217) forbid conducting parametric analyses on ordinal data as "intervals between values cannot be presumed equal," empirical studies of the robustness of tests of central tendency, including t-tests and ANOVA, find that parametric tests on differences in means for sample sizes greater than 5 show robustness with respect to ordinality and non-normality (Boneau 1960; Norman 2010; Pearson 1931). In addition, Wilcoxon signed-rank tests were also conducted with similar results, enhancing confidence in the parametric test outcomes.

[8] 129 players completed playing the game; 105 players completed all the after-survey questions presented afterward.

Table 6.1 Main Effects of Digital Game on Key Dependent Variables

Dependent Variables (n = 129)	Digital Game Δ Post-Pre	Cohen's *d*
Perceived knowledge of climate change (Likert scale: 1–3)	**0.049*** (0.023)	0.195
Perceived personal threat from climate change (Likert scale: 1–5)	−0.115 (0.07)	0.149
Concern for local climate risks (Likert scale: 1–5)	**0.175***** (0.081)	0.34
Willingness to pay higher taxes for climate adaptation (Likert scale: 1–5)	**0.058*** (0.032)	0.167

Note: Standard errors in parentheses. All mean comparisons significant at *p < 0.5 (bold face), **p < 0.1 (bold face), ***p < 0.01 (bold face). Unequal variances assumed (t-test). Cohen's *d* is a standardized measure of effect size; values between 0.3 and 0.6 are considered "moderate" effect sizes in behavioral sciences (Cohen 1988).

differences in knowledge, affect, or behavior around climate change at the local level before and after playing the digital game, focusing on the health impacts of climate change.

Table 6.1 sums up the main effects of the digital game on key dependent variables. In line with expectations, digital game participants showed statistically significant increases in their perceived knowledge of climate change, concern for local climate risks, and willingness to pay higher taxes for their municipality to plan for the impacts of climate change after playing the digital game when compared to before playing the digital game. Playing the game was not associated with statistically significantly higher levels of perceived personal threats from climate change.

Shifts in knowledge, affect, and behavior by political viewpoint

Table 6.2 shows the effects of the digital game on key dependent variables by political viewpoint. As political orientation is the biggest determinant of belief in climate change (Hornsey et al. 2016) and, furthermore, people are likely to selectively pick data most consistent with their political outlooks (Kahan et al. 2012; Kahan et al. 2013), it is important for public engagement tools in climate planning and policymaking processes to lead to changes not only in the views of moderates and liberals but in the views of conservatives as well (van der Linden et al. 2016).

However, as can be seen in Table 6.2 showing the main effects of the digital game on key dependent variables by political viewpoint, the digital game was not associated with any shifts in perceived knowledge of climate change, concern for local climate risks, or willingness to pay higher taxes for climate adaptation in politically conservative participants. The game showed a slight effect on increasing the willingness of moderates to pay more in taxes for climate adaptation efforts but did not result in any increases in the other dependent variables. The majority of the increases in the perceived knowledge

Table 6.2 Main Effects of Digital Game on Key Dependent Variables by Political Viewpoint

Baseline Means before Digital Game (Average Treatment Effect)	Average (n = 129)	Conservative	Moderate	Liberal
Perceived knowledge of climate change* (1–3)	2.612 **(+0.039)**	2.333 **(+0.0)**	2.2 **(+0.0)**	2.618 **(+0.045)**
Perceived personal threat from climate change (1–5)	4.385 (−0.135)	3.0 (+0.0)	3.4 (+0.0)	4.517 (−0.124)
Concern for local climate risks*** (1–5)	3.942 **(+0.144)**	3.0 **(+0.0)**	3.2 **(+0.0)**	4.045 **(+0.18)**
Willingness to pay higher taxes for climate adaptation* (1–5)	3.875 **(+0.048)**	3.0 **(+0.0)**	3.0 **(+0.2)**	4.011 **(+0.045)**

Note: Dependent variables with asterisks have mean increases significant at $*p < 0.5$, $**p < 0.1$, $***p < 0.01$.

of climate change, concern for local climate risks, and willingness to pay higher taxes actually came from the politically liberal people that participated. Therefore, the digital game was effective in shifting the views of liberal participants who were already more likely to know about climate change, be concerned about local climate impacts, and more willing to pay for climate adaptation. However, the digital game's ability to effect counter-attitudinal transformational learning (like that of the face-to-face role-play simulation) seems to be more limited.

Different types of learning in the digital game

Table 6.3 depicts how game participants ranked the relative amounts of learning in different dimensions of climate adaptation from the role-play simulation. In the first column (Average) of Table 6.3, the average participant said the digital game helped them the most in understanding the viewpoints of different stakeholders in dealing with climate change in their municipality (3.12), then the substantive health impacts of climate change (2.86), and, finally, imagining how stakeholders could reach an agreement on how to proceed with climate adaptation at the local level (2.65). Considering how the digital game is a solitary game played alone at home or through one's mobile phone and that the game lets users step into multiple hypothetical stakeholders' shoes, the findings that participants felt they learned the most about the views of other stakeholders, and the least about reaching an agreement in climate adaptation planning and policymaking processes, seem fairly intuitive.

For the substantive learning about the health impacts of climate change, politically moderate participants said they learned the most, then liberal participants, and finally the conservative participants. For learning about the various views of stakeholders in local climate adaptation planning and policymaking processes, again, moderates said they learned the most and conservatives the least. Finally, for improving people's ability

Table 6.3 Effects of Digital Game on Different Types of Learning by Political Viewpoint

Perceived Learning in Digital Game (1–5)	Average (n = 105)	Conservative	Moderate	Liberal
Health impacts of climate change	2.86	2.33	3	2.92
Views of other stakeholders in local climate adaptation	3.12	2.33	3.6	3.18
Reaching an agreement in local climate adaptation	2.65	2.33	2.8	2.7

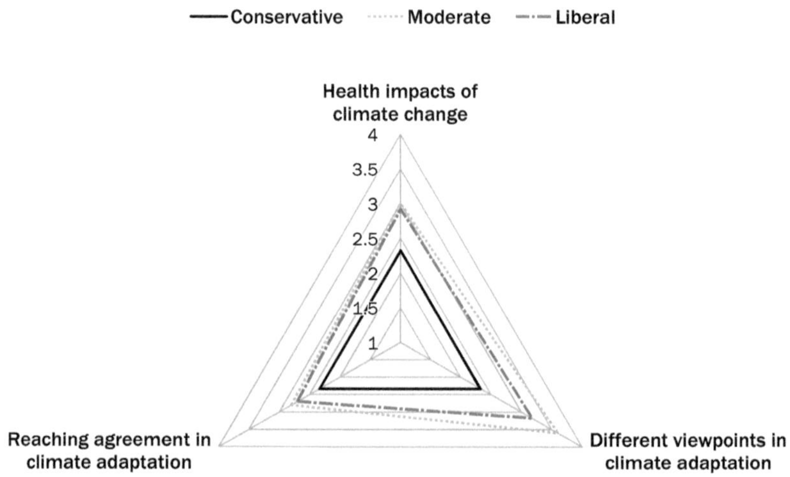

Figure 6.5 Effects of digital game on different types of learning by political viewpoint

to imagine how stakeholders could reach an agreement on how to proceed with climate adaptation at the local level, there is a similar pattern, where moderate participants learned the most, then liberal participants, and finally conservatives. These differences are visualized in Figure 6.5.

Shifts in local climate adaptation policy priorities

Playing the digital game also shifted participants' opinions regarding the policy actions their municipality should take to adapt to climate change. Table 6.4 depicts the shift in the proportions of respondents' prioritization of local climate risks to focus on in their city's climate adaptation efforts, before and after the digital game. The percentages represent the proportion of the group that selected each climate risk as their first priority in their city's climate plans and policies.

There is a modest statistically significant shift in the game players' prioritizing the increased risks of disease and death, going from 16.1 percent to 22.9 percent, with a simultaneous decrease in prioritizing infrastructure damage from 65.3 percent to 60.2 percent and ecosystem degradation from 16.1 percent to 14.4 percent, after the

Table 6.4 Prioritization of Local Climate Risks in Climate Adaptation, Before and After Digital Game

	Before	After
Individual property damage	1.7%	1.7%
Local business disruption	0.9%	0.9%
Increased risks of disease, hospitalization, and death	16.1%	22.9%
Disproportionate impacts to vulnerable populations	0.0%	0.0%
Water, energy, and transportation infrastructure damage	65.3%	60.2%
Watershed, forest, and ecosystem degradation	16.1%	14.4%
Total	100%	

digital game.[9] Again, similarly, this points to the digital game's potential to shift people's thinking about the importance of human resilience to climate change. The digital game emphasizes the health implications of climate change at the local level and, analogous to the emphasis framing effect studied in Chapter 4, induces participants to change their policy preferences to prioritize the reduction of adverse health outcomes from climate change.

Proactive learning

For two of the four characters, a preparedness "tool" provided a link to an external website where the player could directly obtain more information on the specific climate-related health risk. As an example, for Martin, the architect mentioned above with a history of cardiac issues, climate change increases his heat vulnerability. One of the preparedness tools consisted of looking up the relevant real-time local temperature and humidity information online. The heat index factors in humidity and temperature to approximate how one actually feels, as humidity impairs the body's ability to cool itself. As can be seen in Figure 6.6, after the player selects the tool of checking the heat index for Martin, they can actually click on the URL link to go to a popular weather website to check the heat index at the moment for their zip code.

Similarly, for Cora, the child with asthma, changes in air quality due to climate change are expected to increase the number and severity of asthma and allergy episodes. For Cora, one of the three preparedness tools entailed checking the air quality index online before going outdoors or starting any strenuous activities. The link to the external website presented in the explanation for the caretaker action is shown in Figure 6.7.

9 All other marginal proportions are similar before and after the intervention. Results of a McNemar's test show that the marginal proportions are statistically significantly different from each other (Fisher's exact score = 0.000). The McNemar's test is a statistical test used on contingency tables with matched pairs of subjects to determine whether the row and column marginal frequencies are equal. If the Fisher's exact score is low enough ($p < 0.05$), this provides sufficient evidence to reject the null hypothesis and state that the marginal proportions are statistically significantly different from each other.

Figure 6.6 Proactive learning: Checking the heat index

Figure 6.7 Proactive learning: Checking the air quality index

However, a majority (94.31 percent) of the 123 participants did not click on either URL to be taken to a website to check the heat index or the air quality index for their neighborhood when playing the game. Six people clicked on either of the URLs for Cora or Martin, and only one player clicked on both URLs for both characters. Overall, the participants did not engage in more proactive learning through the game to obtain relevant heat and air quality data for their neighborhood.

While this "proactive learning" was not found to be associated with (1) performance on true/false questions at the end of each character phase, (2) different types of learning, or (3) shifts in perceived knowledge of climate change, concern for local climate impacts, willingness to pay higher taxes for climate adaptation, or policy priorities in climate adaptation, this finding demonstrates the challenge of designing a digital game experience to achieve the depth of experiential learning possible in face-to-face environments. While the low barriers to participation and ease of access may allow the game to reach a broader population, the digital game format also allows for less control of a structured learning and facilitation experience.

Themes from citizen input: Connecting game results to real-life situations

Out of the 105 participants who completed the game and the after-survey, 77 people provided answers to the question: "What specific actions would you recommend the city of Cambridge take to manage the risks of climate change?" The participants provided a wide range of opinions on what their city should do to adapt to climate change, but a few key themes emerged.

First, many players argued for the city to prioritize health in climate adaptation planning and policymaking efforts. One player recommended that Cambridge foster "community-wide networking and city policy which has health of people and earth as priority over development." Similarly, another player stated that Cambridge should "continue to improve the health and safety codes of buildings and public infrastructure to make getting around easier and safe for everyone."

Second, many highlighted the importance of further public engagement and education. The most frequent word in all of the inputs provided was "education." One participant asked of their city, "Begin having community discussions hosted at City Hall so people can express concerns, ask questions, and understand what to do or collaborate on solutions." Another game player similarly said, "Provide ongoing education efforts and public forums for discussing and deciding on a range of approaches." In terms of the relevant stakeholder groups and venues, one player said, "Educate religious leaders and doctors so they can talk to their parishioners and patients about climate change, and what actions to take. Also make this game available through school, library, and adult education programs!" A couple of players emphasized the importance of having education efforts tailored for vulnerable populations as well.

Similarly, game players valued information provision and called for more information on sources of vulnerability and existing resources in addressing vulnerability in their municipality. Players were particularly intent on having more information about emergency response resources at hand, such as the locations of cooling centers for extreme heat events. In addition, one player asserted that the city should "keep the issue visible to citizens and provide contact information for government and private groups working to prevent climate change." Another player said that the city should "give residents a list of what they can do personally to reduce climate change." Finally, one participant stated the importance of not only having more information but also making it more accessible: "Have more information available in languages other than English; make all information more visible so that more people are aware of the dangers to the specific groups shown here [in the game]."

Third, a significant number of participants emphasized the importance of renewable energy. Many people stated that the municipality should shift to 100 percent renewable energy by pursuing zero-emissions public transit and the conversion of city fleets to electric vehicles. In the meantime, many players also emphasized the importance of energy efficiency and conservation. Again, similar to the role-play simulation participant group (in Chapter 5), the digital game participants showed much more knowledge of and interest in climate mitigation activities than that of climate adaptation.

A couple of players argued that the city should do more for emergency preparedness. In particular, many concerns were voiced about potential coastal storms and flooding. For instance, one game player said, "I have nowhere to go in an extreme wind event and no basement. We need storm shelters. [We also need] events to make it easy to understand how to make emergency supplies (what to store them in and where in small spaces)." Another participant stated, "Organize evacuation routes and plan for people who don't drive, especially. Plan for evacuation." After engaging with the challenges of managing the health risks of climate change at the municipal level, both the digital game and face-to-face role-play simulation participants (see Chapter 5) emphasized prioritizing emergency preparedness and enhancing the municipality's capacity to prepare for and respond to acute events.

Out of the locally relevant climate impacts in Cambridge, flooding seemed to be the most salient risk in game players' minds, as there were several actions presented that targeted flooding specifically, such as "I think restoring flood plains and planting more vegetation is crucial to mitigate the inevitable flooding that is coming." Another game player stated, "Organize an effort to install the other pump at Amelia Earhart Dam and raise the dam height 6–10 feet." Some people did mention extreme heat, among other extreme events. They advised the city to reduce the urban heat island effect through building more parks and increasing trees on streets to provide shade and cooling.

In summary, the digital game participants presented a variety of suggestions to the city of Cambridge on how to respond to the health risks associated with climate change. Local governments, even at traditional public meetings, rarely get a chance to ask citizens about specific steps they can take to tackle abstract, large-scale policy issues like climate change (Susskind et al. 2015). Therefore, digital games can be used not only to facilitate learning about a policy issue but also to elicit opinions or advice from game players on matters that affect citizens' well-being and assets.

Inside the digital game experience

Finally, many game players expressed their opinions about the digital game during and after the game play period. Comments received via email and social media and post-game interviews[10] conducted with game participants elucidated how the digital game helped them learn about the likely health impacts of climate change within their community.

Many participants remarked on how they identified and empathized with certain characters and thought about the disproportionate impacts of climate change on more

[10] n = 2.

vulnerable members of the community. One participant remarked on the merits of having a range of characters to delve into: "I really liked having the different characters, with a mix of older people and younger people. I think it's important to feel connected to the character, and I readily feel connected to characters similar to me, so it's a good idea to present different characters so that lots of people can connect with at least one character." Another game player commented on one character that stood out, saying, "I found the veteran who lost his legs in a wheelchair compelling without being overly shocking." Another game player remarked on how seeing a child getting sick on screen made her think more about climate change: "My favorite was Cora (child with respiratory issues). I found myself thinking, 'Oh no, her asthma will get worse!'" And finally, one participant commented, "My heart went out to Sofia (the recent immigrant)." She went on to observe that this character made her more aware of certain groups in her city, saying, "And I'm very glad you included the immigrant, or the person for whom English is not their first language, because it's something I hadn't really thought about."

In addition, the game induced players to think about issues they normally did not spend much time or effort on. One participant remarked on how until she had played the game she had never thought about how climate change could affect her health: "I had mostly just thought about flooding, and weather conditions, and not much about health impacts [from climate change]." According to another player, the game made climate change more concrete. He said, "I think the game is effective because it makes climate change personally relevant. We're so used to bad news, that we're not bothered by it, but if it's personal, I think we pay a little more attention." Finally, one game player remarked on how the game made her reexamine heat as a potential threat to health: "I was surprised by the heat problems depicted in the game. I guess I didn't really consider that fact that a lot of people don't have air conditioners."

However, not everyone found the digital game experience to be technically smooth or engaging. Some participants pointed out that certain elements of the digital game felt less than real. One player commented that they found the game to be passive, as "you click on people and items and are fed text." Another participant also remarked that he found the game to be more like "an online brochure."

Many players had issues with starting the game or selecting radio buttons to record their responses for the before- or after-survey questions. Several players also reported confusion about next steps to take in the process of playing the digital game. Lastly, about a dozen players encountered browser and mobile phone compatibility issues during the game play period. These instances point to the importance of providing optional instructions for playing the game, incorporating a troubleshooting element into the game to respond to participants with technical issues, and, overall, further reducing technical barriers to playing the digital game.

Conclusion

The digital game used in the *Frames and Games* research project provided personal accounts of characters representing four distinct subpopulations more vulnerable to the health impacts of climate change due to age, occupation, English proficiency, and/or existing health conditions. The game also highlighted actionable steps that players can take in daily life to alleviate these health risks, either personally or in a caretaker role.

Playing the digital game was associated with increases in perceived knowledge of climate change, concern for local climate impacts, and willingness to pay higher taxes for local climate adaptation efforts. In addition, learning about and engaging with the health considerations of climate change also induced game participants to change their policy preferences to prioritize emergency preparedness and response policies. The average game players said the digital game helped them the most in understanding the viewpoints of different stakeholders in dealing with climate change in their municipality, then the substantive health impacts of climate change, and, finally, imagining how local stakeholders could reach an agreement on how to proceed with climate adaptation. Finally, the digital game participants presented a variety of suggestions to their local government on how to respond to the health risks associated with climate change, proving how digital games can be used not only to facilitate learning about a policy issue but also to elicit opinions or advice from game players on matters that affect people's well-being and assets.

Chapter 7

ROLE-PLAY SIMULATIONS VERSUS DIGITAL GAMES

This chapter juxtaposes the results of the face-to-face role-play simulation (Chapter 5) and the digital game (Chapter 6) to compare the two serious games and their impacts on climate adaptation planning and policymaking processes at the local level.

Demographics

As seen in Table 7.1, the role-play simulation and digital game sample demographics are fairly similar. The role-play simulation sample has a more balanced gender breakdown and is slightly younger in age. However, the digital game reached a broader audience in terms of political viewpoints, with a bigger share of political conservatives and moderates. In addition, the digital game sample is more balanced in terms of household income. Both samples are similarly well educated and work in similar industries.

Attitudinal Changes

While both the face-to-face role-play simulation and the digital game were associated with increases in perceived knowledge of climate change and concern for local climate risks, the face-to-face role-play simulation effected larger shifts in its participants (knowledge of climate change: 0.073 vs. 0.049; concern for local climate risks: 0.259 vs. 0.175). The results are presented in Table 7.2.

Attitudinal changes by political viewpoint

More importantly, there were profound differences in the two games' outcomes across the political spectrum. The face-to-face role-play simulation had the largest effects in bolstering one's perceived knowledge of climate change in political conservatives compared to moderates or liberals. Similarly, the role-play simulation effected a bigger change in amplifying one's concern for local climate risks in political conservatives than in moderates. The relevant numbers are presented in Table 7.3 (numbers underlined). The significance of the face-to-face role-play simulation effecting shifts in the thoughts, feelings, and behavioral intentions of not only moderates and liberals but also, and especially, conservatives cannot be overstated.

Table 7.1 Role-Play Simulation versus Digital Game Demographics

Variables		Role-Play Simulation (n = 112)	Digital Game (n = 105)
Gender	Male	47.6%	28.6%
	Female	52.4%	70.5%
Age	19 or under	1.0%	7.7%
	20–29	13.5%	
	30–39	18.3%	14.3%
	40–49	15.4%	19.1%
	50–59	21.2%	19.1%
	60+	30.8%	40.0%
Political Viewpoint	Conservative	11.5%	21.9%
	Moderate	7.7%	19.1%
	Liberal	80.8%	59.1%
Education	High school graduate (or equivalent)	2.9%	1.0%
	Some college or associate's degree (AA)	2.9%	2.9%
	Bachelor's degree (BA, BS, AB, etc.)	24.0%	26.0%
	Master's degree (MA, MSc)	70.2%	88.5%
	Professional or doctoral degree (MD, PhD, etc.)		
Industry	Agriculture, forestry, fishing and hunting, mining	0%	1.0%
	Construction	3.1%	1.9%
	Manufacturing	1.0%	0.0%
	Wholesale trade	0%	0.0%
	Retail trade	0%	1.0%
	Transportation, warehousing, and utilities	2.1%	10.0%
	Management and administrative services	3.1%	9.7%
	Information, finance, insurance, real estate	13.5%	8.7%
	Educational services and healthcare	36.5%	33.0%
	Arts, entertainment, recreation, hospitality	3.1%	7.8%
	Public administration	5.2%	8.7%
	Other	32.3%	28.2%

In contrast, the digital game had *no effect* in increasing perceived knowledge of climate change, concern for local climate risks, or willingness to pay higher taxes for climate adaptation in political conservatives. In political moderates, the digital game led to an increase in willingness to pay higher taxes. In other words, the shifts in these factors from playing the digital game primarily took place among liberal participants. These numbers are presented in more detail in Table 7.4 (emphasis on underlined figures). Therefore, while the face-to-face role-play simulation cultivated counter-attitudinal transformative learning, the digital game mostly fostered shifts in the cognition, affect, and behavior of politically liberal participants already likely to know more about, be more concerned about, and more likely to support local climate policies in the first place.

What caused these differences in the games' efficacy in shifting participants' views? The interviews conducted with the face-to-face role-play simulation participants pointed

Table 7.2 Main Effects of RPS versus Digital Game on Key Dependent Variables

Dependent Variables	Role-Play Simulation	Cohen's *d*	Digital Game	Cohen's *d*
	Δ Post-Pre		Δ Post-Pre	
Perceived knowledge of climate change (1–3)	**0.073*** (0.031)	0.226	**0.049*** (0.023)	0.195
Perceived personal threat from climate change (1–5)	0.008 (0.051)	0.016	−0.115 (0.07)	0.149
Concern for local climate risks (1–5)	**0.259**** (0.099)	0.247	**0.175***** (0.081)	0.34
Willingness to pay higher taxes for climate adaptation (1–5)	−0.083 (0.076)	0.104	**0.058*** (0.032)	0.167

Note: Standard errors in parentheses. All mean comparisons significant at *p < 0.5 (bold face), **p < 0.1 (bold face), ***p < 0.01 (bold face). Unequal variances assumed (t-test). Cohen's *d* is a standardized measure of effect size; values between 0.3 and 0.6 are considered "moderate" effect sizes in behavioral sciences (Cohen 1988).

Table 7.3 Main Effects of RPS on Key Dependent Variables by Political Viewpoint

Baseline Pre-RPS Means (Average Treatment Effect)	Average (n = 112)	Conservative	Moderate	Liberal
Perceived knowledge of climate change* (1–3)	2.624 (+0.073)	2.2 **(+0.2)**	2.57 **(+0.143)**	2.64 **(+0.07)**
Perceived personal threat from climate change (1–5)	4.446 (+0.009)	3.6 (+0.0)	4.14 (+0.143)	4.55 (+0.022)
Concern for local climate risks** (1–5)	3.901 (+0.259)	3.2 **(+0.2)**	3.43 **(+0.143)**	4.0 **(+0.281)**
Willingness to pay higher taxes for climate adaptation (1–5)	3.862 (−0.083)	3.2 (−0.4)	3.0 (−0.143)	3.99 (−0.081)

Note: Dependent variables with asterisks have mean increases significant at *p < 0.5, **p < 0.1, ***p < 0.01.

to the debrief sessions, where people applied what happened in the game to their actual lives as citizens in their municipality, as a primary enabling factor in people shifting their views on climate change at the local level. In contrast, the digital game had a solitary structure, where the game players learned about each of the four characters of the game on their mobile devices or through their personal computers, without a chance to process the results of the game in concert with other game players. Therefore, the design of the game experience, particularly in whether the game players undertake a collective debriefing session after playing the game, is likely to be linked to the extent of counter-attitudinal learning among game participants.

Table 7.4 Main Effects of Digital Game on Key Dependent Variables by Political Viewpoint

Baseline Pre–Digital Game Means (Average Treatment Effect)	Average (n = 129)	Conservative	Moderate	Liberal
Perceived knowledge of climate change* (1–3)	2.612 (+0.039)	2.333 **(+0.0)**	2.2 **(+0.0)**	2.618 **(+0.045)**
Perceived personal threat from climate change (1–5)	4.385 (−0.135)	3.0 (+0.0)	3.4 (+0.0)	4.517 (−0.124)
Concern for local climate risks*** (1–5)	3.942 (+0.144)	3.0 **(+0.0)**	3.2 **(+0.0)**	4.045 **(+0.18)**
Willingness to pay higher taxes for climate adaptation* (1–5)	3.875 (+0.048)	3.0 **(+0.0)**	3.0 **(+0.2)**	4.011 **(+0.045)**

Note: Dependent variables with asterisks have mean increases significant at *$p < 0.5$, **$p < 0.1$, ***$p < 0.01$.

Different Types of Learning

In addition, the role-play simulation and digital game show different relative strengths in participants' perceived learning about climate-health linkages and municipal adaptation planning and policymaking processes. The digital game was more effective in participants' substantive learning about the health impacts of climate change, while the role-play simulation was more effective in enabling people to envision possible future adaptation pathways and learn to reach an agreement with other stakeholders in local climate adaptation efforts.

In Tables 7.5 and 7.6, compared to that of the face-to-face role-play simulation participants, the participants of the digital game said they learned more about the health impacts of climate change. Counter to the expected results, the digital game players had higher levels of perceived learning about the views of other stakeholders in municipal climate adaptation planning and policymaking processes. This could be from the design of the digital game, in which players learned about a wider range of subpopulations vulnerable to the health impacts of climate change through the four characters available to the participant, whereas in the face-to-face role-play simulation, the players assumed one role for the duration of the game session. On the other hand, compared to the group of people that played the digital game, the face-to-face role-play simulation participants said that the simulation helped them more in imagining how their own city could proceed with climate adaptation in the future through collaborating with other stakeholders in climate adaptation efforts at the local level.

In conclusion, the two games possess different strengths in the different dimensions of learning about climate adaptation at the local level: while the digital game helped participants gain a better understanding of the health implications of climate change within their own city, particularly in how certain groups of people are more vulnerable to these likely impacts because of age, gender, mobility, preexisting health conditions, or English proficiency, the face-to-face role-play simulation helped participants imagine a

Table 7.5 Effects of RPS on Different Types of Learning by Political Viewpoint

Perceived Learning in Role-Play Simulation (1–5)	Average (n = 112)	Conservative	Moderate	Liberal
Health impacts of climate change	2.79	2.33	2.75	2.83
Views of other stakeholders in climate adaptation	2.83	2.33	3.25	2.83
Reaching an agreement in climate adaptation	2.8	2	3	2.83

Table 7.6 Effects of Digital Game on Different Types of Learning by Political Viewpoint

Perceived Learning in Digital Game (1–5)	Average (n = 105)	Conservative	Moderate	Liberal
Health impacts of climate change	2.86	2.33	3	2.92
Views of other stakeholders in climate adaptation	3.12	2.33	3.6	3.18
Reaching an agreement in climate adaptation	2.65	2.33	2.8	2.7

way forward through multi-stakeholder collaboration in municipal climate adaptation processes.

Relative Advantages and Disadvantages of Game Format

The *Frames and Games* project probed the inevitable tradeoff that many public engagement efforts face between breadth and depth. As mentioned in Chapter 3, Farrington and Bebbington's (1993) axes situate different forms of participation, with depth as one axis indicating degrees of involvement and breadth, representing inclusion and exclusion as the other, indicating the tradeoff that many public engagement efforts face in involving citizens in different planning and policymaking processes.

For example, face-to-face public engagement efforts are high on the depth axis and may have more potential for transformative learning (Rumore et al. 2016; Susskind et al. 2015) but may find it difficult to secure broad-based participation beyond the "usual suspects" and score equally high on the inclusion axis (Beierle 1999; Konisky 2001). In contrast, public engagement using digital technologies pose lower costs and barriers to entry to participants, indicating more potential for breadth, but questions remain on the level of depth possible with online forms of public engagement (Best and Krueger 2005; Evans-Cowley & Hollander 2010; Gordon and Baldwin-Philippi, 2014).

As can be seen in the comparisons of the game formats above, the face-to-face role-play simulation and digital game are considerably different in the benefits and trade-offs they offer to cities as public engagement tools in climate planning and policymaking processes. This goes hand in hand with research to date covering the wide variety of climate games in different formats and the relative strengths and weaknesses of each format (Wu Lee 2015).

Table 7.7 Summary and Comparison of Face-to-Face RPSs and Digital Games

Game Format	Key features of Featured Games	Pros	Cons	Goals and Outcomes
Face-to-Face Role-Play Simulation	Role-playing mock decision-making process; Facilitated post-game debrief	Flexible and adaptable experience; Fosters counter-attitudinal transformative learning	Higher costs (game development and deployment); Facilitation requirement a potential barrier	Facilitated learning with debriefing; In-depth qualitative assessment
Digital Game	Computer-based character game with actionable steps for each character	Consistent and scalable experience; Lower costs (game development and deployment); Fosters substantive learning about climate change	Computer or mobile phone and technical literacy requirement a potential barrier	In-game assessment

Adapted from Wu and Lee (2015, 416).

Table 7.7 summarizes the comparisons drawn between the two game formats used in the *Frames and Games* research project. A key takeaway gleaned from analyzing the outcomes of the face-to-face role-play simulation and digital game is that both game formats have different comparative strengths depending on intended learning outcomes. The face-to-face role-play simulation has a definite advantage in being flexible and adaptable to each group of participants attending, particularly in how each debrief session is tailored to target specific learning goals. In addition, through the debriefing sessions structured for participation reflection and peer learning, the face-to-face role-play simulation achieves cross-attitudinal transformative learning, leading to increases in perceived urgency of climate risks and support for local climate adaptation, potentially counter to one's own political viewpoints.

In contrast, the digital game's relative strength is in the substantive learning about local climate risks and the health implications of climate change, particularly in how climate change most affects the health of already vulnerable populations in any given locality. Digital games allow for in-game assessment of learning, such as questions or points, that gauge players' learning regarding climate change. The digital game, with its true/false questions to check learning about climate change and resulting health impacts and proactive learning mechanisms about these phenomena in one's own neighborhood, resulted in game players saying they had learned more about substantive areas of climate change and public health after playing the digital version, when compared to the responses given by people who had participated in the face-to-face role-play simulation.

In addition, digital games are easier to deploy at scale. Face-to-face events are one of the most resource-intensive forms of public engagement in terms of money, staffing, and time. In contrast, digital games have upfront costs of game development but relatively minimal launch and maintenance costs, making it much easier to reach larger audiences. Digital games also offer a more consistent learning experience and faster learning assessments with electronic player data collection and analysis (Bellotti et al. 2013; Wu and Lee 2015). However, the results suggests that the digital game-based learning may be confined to low-level learning (Argyris and Schon 1978; Wu and Lee 2015), as shown by the smaller shifts in concern about climate change and commitment to climate adaptation action at the local level.

Therefore, cities should harness the relative strengths of both face-to-face and digital games in furthering public engagement with climate change and strengthening the human resilience aspects of their climate planning and policymaking processes.

Chapter 8

DISCUSSION AND POLICY RECOMMENDATIONS

Framing Climate Change as a Public Health Issue

Urban climate adaptation is often framed as protecting a city's physical assets from hazards and disasters. As a result, critiques have pointed out that adaptation plans rarely incorporate equity or social vulnerability, and calls have emerged for climate adaptation plans and policies to focus on building not just physical resilience but also human resilience to climate change (IPCC 2014; Hughes 2015).

In addition, cities have found that most publics in the United States do not find adapting to the changing climate to be a political priority (Howe et al. 2015), highlighting the importance of communicating the need for climate adaptation to elected officials and the public (Carmin et al. 2012; Howe et al. 2015; Shi et al. 2015). However, current public engagement practices are likely to be inadequate in fostering the technical and political learning necessary to collectively manage the climate risks that our cities face (Argyris and Schon 1978; de Suarez et al. 2012; Susskind 2001).

The *Frames and Games* project's survey experiment results show that framing climate change as a public health issue increases perceived knowledge of climate change, concern for local climate risks, and confidence in the local government's capacity to effectively manage climate risks. This indicates that cities have much to gain from framing climate change as a public health issue, as it boosts perceptions about the severity of the problem and builds commitment to policy action. In other words, a public health frame can be more powerful than an environment frame of climate change in motivating people to support climate adaptation plans and policies at the local level.

The public health frame of climate change also accentuates the potential of climate change to increase the prevalence of infectious diseases and other health problems that are already perceived as important, especially among the most vulnerable populations. Furthermore, according to interviews conducted as part of the project, the public health frame also heightens the geographic, temporal, and social proximity of climate impacts, corroborating previous research on the efficacy of the frame as a public education and engagement tool for climate planning and policymaking efforts (Frumkin et al. 2008; Maibach et al. 2010).

Therefore, cities should frame climate change as a public health issue. The proposed normative benefits of a stronger emphasis on the public health impacts of climate change revolve around alleviating the lack of attention of current climate adaptation efforts to equity and social vulnerability (Frumkin et al. 2008; Watts et al. 2015; WHO 2013). On the pragmatic side of things, highlighting climate-related health risks allows climate change to transcend political schisms between climate advocates and dismissives or deniers (Maibach et al. 2010; Myers et al. 2012; Nisbet 2009).

The *Frames and Games* research project developed and tested a face-to-face role-play simulation and a digital game that cities can use to operationalize the public health frame of climate change, engage publics with in climate policymaking processes, and, furthermore, strengthen collective problem-solving capacities in responding to increasing threats to public health and well-being from climate change.

Face-to-Face Role-Play Simulations

Role-play simulations, or serious games that engage participants in a face-to-face mock decision-making process, are effective tools for learning and mutual deliberation about how a city should deal with the adverse health impacts of climate change. After participating in the role-play simulation designed for the *Frames and Games* research project, people showed increases in their perceived knowledge of climate change, concern about local climate risks, and confidence in effective local climate adaptation action.

In particular, it is important to note that with the role-play simulation, the conservative participants (versus moderate or liberal) showed the largest increases in knowledge of climate change, concern about local climate impacts, and confidence in the city's effectiveness in local climate adaptation. In light of the political polarization of climate change as a policy issue in the United States, the face-to-face role-play simulation's potential to catalyze counter-attitudinal transformative policy learning cannot be overstated.

Furthermore, the role-play simulation also helped clarify the range of policy preferences among citizens in the actions their city should take to adapt to climate change. After playing the role-play simulation, more people said they would prioritize the health risks of climate change in local climate adaptation efforts. In addition, qualitative data collected during and after the role-play simulation point to the efficacy of role-play simulations in facilitating learning in process-based knowledge of how planning and policymaking works in a municipality, fostering perspective-taking of different stakeholders in climate adaptation efforts at the local level and strengthening the place identity and attachment of participants.

Digital Games

The *Frames and Games* project's digital game was designed to depict personal accounts of characters representing subpopulations more vulnerable to the health impacts of climate change, along with actionable steps that the game player can take in daily life to alleviate these health risks. Playing the digital game boosted perceived knowledge of climate change, concern about local climate impacts, and willingness to pay higher taxes for local climate adaptation. In addition, more people chose the increased risks of hospitalization, disease, and death as their first priority to mitigate in local climate adaptation policies after playing the digital game. Finally, various in-game mechanics were deployed in the digital game to generate nuanced assessments of learning outcomes and better measure game participants' learning, such as true/false questions and proactive learning tools.

In contrast to the face-to-face role-play simulation, there were no attitudinal changes in politically conservative citizens that played the digital game. Instead, a majority of the increases in knowledge of, concern about, and policy support for local climate adaptation from playing the digital game took place among liberal participants, who were already more likely to know about climate change, be concerned about, and be committed to climate action before playing the game.

In addition, the digital game had comparative advantages in facilitating substantive learning about the health impacts of climate change and the different groups of people particularly vulnerable to climate-related health risks in a specific municipality, in comparison to learning about policymaking and planning processes or how to collectively plan for climate change.

Face-to-Face Role-Play Simulations versus Digital Games

A key takeaway from the *Frames and Games* project, gleaned from analyzing the outcomes of the face-to-face role-play simulation and digital game, is that the two serious game formats have different strengths in different contexts.

The face-to-face role-play simulation has a definite advantage in being flexible and adaptable to each group of participants, particularly in how each debrief session is tailored to target specific learning goals. In addition, through structured reflection and deliberation in the debrief sessions, the face-to-face role-play simulation achieves cross-attitudinal transformative learning, leading to increases in the perceived urgency of climate risks and support for local climate adaptation, potentially counter to one's own political viewpoints.

In contrast, the digital game offers a more consistent and scalable learning experience. Its relative strength is in the substantive learning about local climate risks and the health implications of climate change, particularly in how climate change most affects the health of already vulnerable populations in any given locality. In addition, digital games can assess players' learning about climate change through questions or other learning evaluation mechanisms within the game itself. The digital game used in this book project, with its true/false questions to check learning about climate-health linkages and proactive learning mechanisms about these phenomena in one's own neighborhood, resulted in game players saying they had learned more about the science of climate change and public health after playing the digital version when compared to the people who had participated in the face-to-face role-play simulation. However, the results from the *Frames and Games* project suggest that digital games may only induce low-level learning, as can be seen in the smaller shifts in concern about climate change manifesting at the local level.

Table 8.1 summarizes the comparisons drawn between the two game formats used in this research project.

Table 8.1 Comparison of Face-to-Face RPSs and Digital Games

Game Format	Key Features of Featured Games	Pros	Cons	Goals and Outcomes
Face-to-Face Role-Play Simulation	Role-playing mock decision-making process; Facilitated post-game debrief	Flexible and adaptable experience; Fosters counter-attitudinal transformative learning	Higher costs (game development and deployment); Facilitation requirement a potential barrier	Facilitated learning with debriefing; In-depth qualitative assessment
Digital Game	Computer-based character game with actionable steps for each character	Consistent and scalable experience; Lower costs (game development and deployment); Fosters substantive learning about climate change	Computer or mobile phone and technical literacy requirement a potential barrier	In-game assessment

Adapted from Wu and Lee (2015, 416).

Policy Recommendations

This section examines the implications of the research findings from the *Frames and Games* project and provides recommendations to cities across the United States to strengthen the focus on building human resilience in their climate planning and policymaking processes.

First, cities should actively frame climate change as an issue impacting citizens' health and well-being today. The *Frames and Games* project's survey experiments showed that framing climate change as a public health issue, in comparison to framing it as an environmental issue, results in increased perceived knowledge, concern about local climate risks, and confidence in the city's ability to tackle climate change through its climate adaptation planning and policymaking processes.

Furthermore, the framing of climate change not only affects how climate adaptation is perceived relative to other municipal goals (Shaw et al. 2014) but also if and how adaptation plans are implemented (Dupuis and Knoepfel 2013). While assessing climate risks and vulnerabilities and identifying adaptation options are generally thought of discrete steps, how a city defines the problem impacts the solution space (Rittel and Webber 1973). Accordingly, problem identification impacts how the city will go about enhancing the awareness of individuals, organizations, and institutions about climate change as well (IPCC 2014, ch. 16).

More specifically, climate-related health risks are often not presented in current local climate adaptation plans and policies, and subsequently, most climate adaptation plans are not aligned with the end goal of mitigating the adverse health impacts of climate change. Climate plans tend to merely acknowledge that climate change will impact public health and fall short of delineating location-specific climate and health linkages and, furthermore, articulating localized strategies to manage the health risks of climate change (Sailor et al. 2016; CDPH 2012). In parallel, most key stakeholders and publics are unaware of and lack a substantial understanding of how climate change affects people's health and well-being in their cities (Maibach et al. 2015).

Therefore, cities should leverage the public health frame of climate change as a conversation starter and subsequently strengthen local climate-health linkages and identify strategies to manage climate-related health risks in their city's climate adaptation plans and policies. As one activist working with vulnerable populations in their city described this process, "When you go to communities and start talking about the things that concern them, issues like asthma resonate. That affects people very tangibly. It affects more low-income, people of color, poor housing conditions, etc. By starting with what concerns people, and helping to address it immediately, you can also broaden into other areas." Time and again in the interviews conducted for this book, competing policy priorities came up as a barrier to climate adaptation gaining more traction in the Cambridge area. Acknowledging that climate change is often not the leading policy priority of most people, instead, cities should connect climate change to issues such as health or safety first and explain how climate change will affect these other priorities or issues that may be more important to the public. Furthermore, better aligning adaptation plans and policies to the end goal of mitigating adverse climate-related health impacts will strengthen cities' human resilience to climate change.

In doing so, here is a short summary (Table 8.2) of the public health risks of climate change that cities can focus on.

Second, cities should deploy face-to-face role-play simulations to enhance the readiness of citizens and key stakeholders to engage in climate adaptation efforts at the local level. While best practices in designing and implementing role-play simulations are documented extensively (see Susskind et al. 2015), some generalizable recommendations based on designing and implementing a public health-oriented climate adaptation role-play simulation include the following:

1. Secure partnerships with relevant municipal agencies responsible for environment, sustainability, and/or climate change, land-use planning, and public health early on in the process of designing and implementing the role-play simulation.

   In Cambridge, this project benefitted tremendously from the partnership among the MIT Science Impact Collaborative, the Community Development Department and its Director of Environmental Planning, John Bolduc, and the Public Health Department and its Director of Environmental Health, Sam Lipson. Through the *Frames and Games* research project, MIT and the City of Cambridge jointly informed each other's strategies and programs to increase the salience and legitimacy of climate-health information and enhance climate decision-making processes within their city. These collaborations can further catalyze the "mainstreaming" of climate preparedness into the activities of other city agencies as well.

Table 8.2 Human Health Effects of Climate Change

Climate Impacts	Health Impacts	Populations Most Affected
Extreme Heat	Premature death Cardiovascular stress and/or failure Heat-related illnesses, such as heat stroke, heat exhaustion, and kidney stones	Children Elderly Diabetics People with respiratory conditions People with cardiovascular conditions Outdoor workers Low-income populations
Poor Air Quality/ Pollution	Increased asthma, allergies, chronic obstructive pulmonary disease (COPD), and other cardiovascular and respiratory diseases	Children Elderly People with respiratory conditions People with cardiovascular conditions Outdoor workers Low-income populations
Wildfires	Injuries and death from burns and smoke inhalation Eye and respiratory illnesses from air pollution Exacerbation of asthma, allergies, chronic obstructive pulmonary disease (COPD), and other cardiovascular and respiratory diseases Population displacement, loss of home and livelihood	People with respiratory conditions
Extreme Weather Events	Injuries or death from drowning Lack of access to drinking water, water- and food-borne diseases from damage to potable water, wastewater, and irrigation systems Population displacement, loss of home and livelihood	Children Elderly Low-income populations
Increased Average Temperatures	Cardiovascular stress and/or failure Increased number and range of: Vector-borne diseases, such as West Nile virus, malaria Water-borne diseases, such as cholera, *E. coli* Food-borne diseases, such as *salmonella* poisoning Allergies from pollen	Children Elderly Diabetics People with respiratory conditions People with allergies Outdoor workers
All Impacts	Mental health effects, such as depression, anxiety, post-traumatic stress disorder (PTSD)	

Adapted from CDPH (2012); USGCRP (2016); Watts et al. (2015); WHO (2013).

2. Base the role-play simulation on localized information on climate risks and emphasize this fact throughout events to further disseminate this information to the community.

   The (perceived) lack of localized information on relevant climate risks in a city is a significant barrier to furthering climate adaptation efforts. During the face-to-face role-play simulation events, many participants asked questions about the scientific information provided in the role-play simulation. Some even asked if they could take the materials with them after the event. The face-to-face role-play simulation events provide a valuable opportunity for technical learning about the local climate and climate-related health risks people face in their municipalities. Therefore, cities should design their role-play simulations, along with the face-to-face events featuring role-play simulations, to maximize this technical learning and enable citizens to better engage with climate planning and policymaking processes in their municipalities. For instance, staff members or facilitators need to be trained as climate communicators to answer questions about or further explain the technical information presented in the role-play simulation.
3. Structure the role-play simulation, and especially the post-simulation debriefing session, to elicit people's preferences in terms of local climate adaptation planning and policymaking processes.

   In the post-simulation interviews, many participants voiced frustration at the lack of outlets in which to express their views regarding climate change to decision-makers in their cities. The debriefing session, where people discuss their game experiences and step back into their daily lives as citizens of a certain locality, should be structured to ask people for their inputs. Furthermore, the input should be aggregated and sent to the relevant municipal agencies after the game-based public engagement events.

Third, cities should deploy digital games to enable substantive learning of the likely public health impacts of climate change and boost collective recognition of location-specific climate and climate-related health risks. Some recommendations on designing and implementing a public health-oriented climate adaptation digital game include the following:

1. Focus the digital game on climate adaptation, with the recognition that most people currently focus on climate mitigation, in particular, individual climate mitigation activities, that is, reducing greenhouse gas emissions by flying less, eating less meat, and so forth.

   Post-game interviews highlighted how game participants had not thought of how to manage location-specific climate risks or design climate adaptation pathways for their city, when compared to how often they thought of climate mitigation actions they could incorporate into their daily lives. Similarly, game players focused primarily on energy-efficiency measures (climate mitigation) when asked to advise which climate policy actions their city should prioritize. Therefore, the game should emphasize the necessity of climate adaptation and the need for planning and policymaking efforts to address current and future climate impacts.

In addition, climate communication research shows that people sustain knowledge of climate change better when the information is coupled with steps they can take to reduce vulnerabilities and prepare for climate-related events (Pidgeon and Fischhoff 2011). Therefore, the digital game should provide actions people can take to also manage location-specific climate-related health risks.

2. Design the game with the aim of reaching the broadest audience possible by alleviating barriers of computer and/or other digital media proficiency.

   In the *Frames and Games* project, even with adaptive programming and extensive pilot-testing, there were numerous inquiries and complaints regarding the functioning of the digital game. Ensuring browser and mobile phone compatibility is crucial. In order to further alleviate the barriers of computer literacy and other technical skills, future efforts should provide not only optional instructions for playing the game but also real-time technical support during the game-play period.

3. Clarify the aim of in-game assessments before incorporating them into the digital game.

   Digital games offer a plethora of opportunities to collect data during the game experience on a variety of factors. For instance, for the *Frames and Games* project's digital game, data were collected on factors such as the browser used for the game, game duration, order of characters played, and so forth. However, after collecting the data, many factors proved to be of minimal use in explaining learning or differences in learning among the game players. While the exact use of different data is hard to delineate before the design and implementation of the digital game, digital game data collection mechanisms should be designed with a clear idea of what each data point is meant to measure and for what purpose.

4. The possibility of a digital public engagement tool engaging a broader audience does not mean it always does.

   Although digital forms of public engagement have been touted for their potential to alleviate the "usual suspect" problem in public engagement, the *Frames and Games* project's digital game players were not more representative of Cambridge's overall population than the face-to-face role-play simulation participants as expected. The digital game participant sample skewed to older, more politically liberal, and more educated people, but in addition, the digital game participant sample was also more skewed in age than the role-play simulation sample (see the demographics section in Chapter 7 for more details). While the digital game sample was more balanced in political views, the findings of the *Frames and Games* project reinforce the importance of not relying on a medium or technology itself to reach a broader public.

   However, it is also clear from the findings that the digital game players logged on to play the game from more dispersed zip codes and at various times throughout the day. In other words, while digital games may not reach a more representative audience or foster the deliberation and the mutual learning seen at face-to-face events, they provide more opportunities to engage people in less proximate places and at more convenient times. In addition, digital games require less money and time to deploy and scale up. Therefore, digital games and other forms of digital engagement should be viewed as a valuable complement to, rather than a replacement for, face-to-face role-play simulations and more conventional public engagement events.

Fourth, cities should deploy the two serious games used in this book, face-to-face role-play simulations and digital games, but in a specific order to maximize complementarities.

1. The digital game should be used first as a substantive primer introducing the relevant health impacts of climate change likely to arise in a specific city. The digital game is cheaper and easier to scale up than the face-to-face role-play simulation. In addition, digital game outreach efforts can also be utilized toward publicizing upcoming face-to-face events. However, the digital game does have limitations on the scope of learning it can facilitate. As its relative strengths lie in the technical learning about the science of climate change and climate-related health risks, the digital game should be used to set the stage first.
2. Subsequently, cities should deploy face-to-face role-play simulations to enhance their municipality's collective readiness to prepare for climate change. The face-to-face role-play simulation's relative strengths are in allowing participants to envision possible future climate adaptation pathways and to build consensus in climate adaptation planning and policymaking efforts. In particular, the facilitated debriefing session is critical, as it is likely to foster cross-attitudinal transformative shifts in perceived knowledge of climate change, concern about location-specific climate risks, and confidence in the local government to effectively plan for and respond to climate change. Therefore, the debriefing sessions should be structured to catalyze deliberations about climate change and its ramifications for a city across the political spectrum. In addition, the face-to-face role-play simulation also cultivates the place identity and attachment of the people that participate, building upon an important source of strength that local climate adaptation efforts can marshal.

The findings of this book indicate that cities should implement multimodal game-based engagement in climate adaptation planning and policymaking processes. This goes hand in hand with research showing that public engagement is most effective with a variety of participatory tools (Kelly et al. 2012). Different publics will need different ways to learn about climate risks and contribute to climate adaptation action at the local level.

Contributions to Theory and Practice

This book engages with and contributes to three areas of theory and practice.

First, this book aims to explain why and how the health impacts of climate change should be given a more prominent role in climate adaptation efforts. To date, climate adaptation has mostly focused on protecting physical assets from potentially catastrophic climatic changes. While the lack of human vulnerability and equity components in adaptation plans and policies have been critiqued by many (IPCC 2014; Hughes 2015; Watts et al. 2015), this has not yet led to a more productive reframing and operationalization of climate adaptation that situates citizens' health and well-being front and center. The book examines how using a public health frame of climate change boosts perceived urgency of location-specific climate risks and policy commitment to climate adaptation efforts at the local level. The findings from the book can be used to help cities better conduct public conversations about preparing for and managing local climate risks to enhance human resilience to climate change.

Second, this book presents new tools for cities to enhance awareness of and facilitate engagement around climate risk management choices. While the importance of public engagement and collective problem-solving in climate adaptation is widely acknowledged (Hughes 2015; Moser and Pike 2015), there is a relative lack of empirical research analyzing the efficacy of various engagement methods with specific testable hypotheses. The book analyzes the results of the design and implementation of two serious games with a set of common dependent variables to facilitate comparisons of the relative strengths and weaknesses of each game format and advise cities on how to structure multimodal game-based engagement in climate adaptation planning and policymaking processes.

Finally, this book examines the roles of policymakers and practitioners in science-intensive planning and policymaking processes, in particular, through addressing the unique challenges to enhancing public engagement with climate change. The book presents ways for climate and health professionals to bridge expert and public views regarding scientific and technical matters. Furthermore, it examines how policymakers and practitioners can foster collective decision-making capacities among different publics and, ultimately, enable technically sound and politically feasible responses by individuals and communities to adapt to climate change.

Appendix A

ISSUE FRAMING VIGNETTES

Environment Frame

According to Dr. Thomas Lovejoy,[1] a research biologist and former Assistant Secretary for Environmental Affairs at the Smithsonian Institution in Washington, DC, climate change is putting new stresses on important species, which can destabilize ecosystems. "It is now clear that over the next few decades climate change is likely to become a major threat to many of America's plant and animal species," says Lovejoy. Rising temperatures and changes in weather have already contributed to disruptions in ecosystems and species declines in many areas of the country and around the world, according to several new studies.

Health Frame

According to Dr. Howard Frumkin,[2] a medical doctor and former director of the National Center for Environmental Health at the US Centers for Disease Control & Prevention (CDC), climate change is increasing the rates of some diseases and is making extreme weather conditions such as heat waves and severe storms more deadly. "It is now clear that over the next few decades, climate change is likely to become a major threat to the health of Americans," says Frumkin. Increasingly, severe heat waves, other extreme weather events, and the spread of infectious diseases have already contributed to illness and deaths in all regions of the country and around the world, according to several new studies.

1 Adapted versions of vignettes used in Myers et al. (2012).
2 Adapted versions of vignettes used in Myers et al. (2012).

Appendix B

ISSUE FRAMING SURVEY QUESTIONS

Q1 Please answer the following questions. All responses will be kept confidential and compiled in total. Survey results will only be used for academic research purposes.

[Environment/Health Frame Groups: insert vignettes here]

Q2 Please enter your zip code. ____________

Page Break

Q3 How much do you know about global warming or climate change?

- ○ I have never heard of it.
- ○ I know something about it.
- ○ I know a great deal about it.
- ○ Not sure.

Q4 How serious of a threat is climate change to you?

- ○ 1: Not at all serious
- ○ 2: Not very serious
- ○ 3: Somewhat serious
- ○ 4: Very serious
- ○ 5: Extremely serious

Q5 How concerned are you about the possible impacts a changing climate might have on your city or town?

- ○ 1: Not at all concerned
- ○ 2: Not very concerned
- ○ 3: Somewhat concerned
- ○ 4: Very concerned
- ○ 5: Extremely concerned

Q6 How confident are you that your city or town will be able to effectively respond to climate-related risks, despite uncertainty about what the future climate will be like?

- ○ 1: Not at all confident
- ○ 2: Not very confident

- ○ 3: Somewhat confident
- ○ 4: Very confident
- ○ 5: Extremely confident

Q7 How do you feel about your city or town incorporating projections of what the climate might be like in 50 years in everyday planning and infrastructure decisions?

- ○ 1: It is a very bad idea.
- ○ 2: It is a somewhat bad idea.
- ○ 3: It is neither a good nor bad idea.
- ○ 4: It is a somewhat good idea.
- ○ 5: It is a very good idea.

Q8 How willing would you be to pay higher taxes so that your city or town can reduce climate change risks?

- ○ 1: Not at all willing
- ○ 2: Not very willing
- ○ 3: Somewhat willing
- ○ 4: Very willing
- ○ 5: Extremely willing

Q9 Your city or town is looking to reduce its climate risks. Which climate change impact do you think your city or town should prioritize in its policies?

- ○ Individual property damage
- ○ Local business disruption
- ○ Increased risks of disease, hospitalization, and death
- ○ Disproportionate impacts to people less able to prepare for climate impacts
- ○ Water, energy, and transportation infrastructure damage
- ○ Watershed, forest, and ecosystem degradation

Page Break

Q10 What is your gender?

- ○ Male
- ○ Female
- ○ Other: ________________________________

Q11 What is your age?

- ○ 19 or under
- ○ 20–29
- ○ 30–39
- ○ 40–49
- ○ 50–59
- ○ 60+

Q12 How would your describe your political viewpoint?

- ○ Extremely conservative
- ○ Conservative
- ○ Slightly conservative
- ○ Moderate, middle of the road
- ○ Slightly liberal
- ○ Liberal
- ○ Extremely liberal
- ○ Other: ____________________

Q13 Do you belong to any non-profit groups that regularly advocate on behalf of environmental conservation and/or protection? (Please select all that apply.)

- ☐ No
- ☐ Yes, a national group (like The Nature Conservatory or National Aubudon Society, etc.)
- ☐ Yes, a local group (like a watershed alliance or local conservation committee, etc.)
- ☐ Yes, other: ____________________

Q14 What industry most accurately describes your primary field?

- ○ Agriculture, forestry, fishing and hunting, mining
- ○ Construction
- ○ Manufacturing
- ○ Wholesale trade
- ○ Retail trade
- ○ Transportation, warehousing, and utilities
- ○ Finance, insurance
- ○ Real estate
- ○ Management and administrative services
- ○ Education
- ○ Healthcare and social assistance
- ○ Arts, entertainment, recreation, hospitality
- ○ Public administration
- ○ Other: ____________________

Q15 What is the highest level of education you have completed?

- ○ No formal schooling completed
- ○ High school graduate (or equivalent)
- ○ Some college or associate's degree (AA, etc.)
- ○ Bachelor's degree (BA, BS, AB, etc.)
- ○ Master's degree (MA, MSc)
- ○ Professional or doctoral degree (MD, JD, PhD, etc.)
- ○ Other: ____________________

Q16 Could you estimate your household's total income for the calendar year of 2016? (Optional)

- ○ Less than $14,999
- ○ $15,000 to $24,999
- ○ $25,000 to $34,999
- ○ $35,000 to $49,999
- ○ $50,000 to $74,999
- ○ $75,000 to $99,999
- ○ $100,000 to $149,999
- ○ $150,000 or more

Appendix C

ROLE-PLAY SIMULATION BEFORE-SURVEY QUESTIONS

Q1.1 Thank you for your interest in the upcoming workshop. Please provide the following information to reserve your spot. We will also email you with more details and information as the workshop approaches. We will not use your contact information for any other purposes.

Q1.2 Please write your name.

First Name ______________
Last Name ______________

Q1.3 What is your preferred email address? We will be sending you workshop details to this address.

- ____________________________

Q1.4 Do you live and/or work in Cambridge?

○ I live in Cambridge.
○ I work in Cambridge.
○ I both live and work in Cambridge.
○ I neither live nor work in Cambridge.

Please answer the following questions. All responses will be kept confidential and compiled in total. Survey results will only be used for academic research purposes.

Q2.1 How much do you know about global warming or climate change?

○ I have never heard of it.
○ I know something about it.
○ I know a great deal about it.
○ Not sure.

Q2.2 How serious of a threat is climate change to you?

○ 1: Not at all serious
○ 2: Not very serious
○ 3: Somewhat serious
○ 4: Very serious
○ 5: Extremely serious

Q2.3 How concerned are you about the possible impacts a changing climate might have on the city of Cambridge?

- ○ 1: Not at all concerned
- ○ 2: Not very concerned
- ○ 3: Somewhat concerned
- ○ 4: Very concerned
- ○ 5: Extremely concerned

Q2.4 How confident are you that Cambridge will be able to effectively respond to climate-related risks, despite uncertainty about what the future climate will be like?

- ○ 1: Not at all confident
- ○ 2: Not very confident
- ○ 3: Somewhat confident
- ○ 4: Very confident
- ○ 5: Extremely confident

Q2.5 How do you feel about the city of Cambridge incorporating projections of what the climate might be like in 50 years in everyday planning and infrastructure decisions?

- ○ 1: It is a very bad idea.
- ○ 2: It is a somewhat bad idea.
- ○ 3: It is neither a good nor bad idea.
- ○ 4: It is a somewhat good idea.
- ○ 5: It is a very good idea.

Q2.6 How willing would you be to pay higher taxes so that Cambridge can reduce climate change risks?

- ○ 1: Not at all willing
- ○ 2: Not very willing
- ○ 3: Somewhat willing
- ○ 4: Very willing
- ○ 5: Extremely willing

Q2.7 The city of Cambridge is looking to reduce its climate risks. Which climate change impact do you think Cambridge should prioritize in its policies?

- ○ Individual property damage
- ○ Local business disruption
- ○ Increased risks of disease, hospitalization, and death
- ○ Disproportionate impacts to people less able to prepare for climate impacts
- ○ Water, energy, and transportation infrastructure damage
- ○ Watershed, forest, and ecosystem degradation

Q2.8 Which policy option should the city of Cambridge prioritize to reduce climate-related risks?

- ○ Information provision
- ○ Emergency preparedness
- ○ Resource allocation
- ○ Rules and regulations
- ○ Prevention

Q2.9 What industry most accurately describes your primary field?

- ○ Agriculture, forestry, fishing and hunting, mining
- ○ Construction
- ○ Manufacturing
- ○ Wholesale trade
- ○ Retail trade
- ○ Transportation, warehousing, and utilities
- ○ Finance, insurance
- ○ Real estate
- ○ Management and administrative services
- ○ Education
- ○ Healthcare and social assistance
- ○ Arts, entertainment, recreation, hospitality
- ○ Public administration
- ○ Other: ____________________

Appendix D

ROLE-PLAY SIMULATION AFTER-SURVEY QUESTIONS

1. Please write your name. ________________________
2. Are you a resident of Cambridge? Yes / No
    a. If you are a resident of Cambridge, how long have you lived in the community? ____ years
3. How much do you know about global warming or climate change?
    a. I have never heard of it.
    b. I know something about it.
    c. I know a great deal about it.
    d. Not sure.
4. How serious of a threat is climate change to you?

1	-	2	-	3	-	4	-	5
Not at all serious		*Not very serious*		*Somewhat serious*		*Very serious*		*Extremely serious*

5. How concerned are you about the possible impacts a changing climate might have on the city of Cambridge?

1	-	2	-	3	-	4	-	5
Not at all concerned		*Not very concerned*		*Somewhat concerned*		*Very concerned*		*Extremely concerned*

6. How confident are you that the city of Cambridge will be able to effectively respond to climate-related risks, despite uncertainty about what the future climate will be like?

1	-	2	-	3	-	4	-	5
Not at all confident		*Not very confident*		*Somewhat confident*		*Very confident*		*Extremely confident*

7. How do you feel about the city of Cambridge incorporating projections of what the climate might be like in 50 years in everyday planning and infrastructure decisions?

1	-	2	-	3	-	4	-	5
Very bad idea		*Somewhat bad idea*		*Neither a good nor bad idea*		*Somewhat good idea*		*Very good idea*

8. How willing would you be to pay higher taxes so that Cambridge can reduce climate change risks?

1	-	2	-	3	-	4	-	5
Not at all willing		*Not very willing*		*Somewhat willing*		*Very willing*		*Extremely willing*

9. The city of Cambridge is looking to reduce its climate risks. Which climate change impact do you think the city of Cambridge should prioritize in its policies?
   a. Individual property damage
   b. Local business disruption
   c. Increased risks of disease, hospitalization, and death
   d. Disproportionate impacts to vulnerable populations
   e. Water, energy, and transportation infrastructure damage
   f. Watershed, forest, and ecosystem degradation

10. Which policy option should the city of Cambridge prioritize to reduce climate-related risks?
    a. Information provision
    b. Emergency preparedness
    c. Resource allocation
    d. Rules and regulations
    e. Prevention

Please answer the following questions about the workshop in general.

11. How much did you learn about the health impacts of climate change?

1 -	2 -	3 -	4 -	5
A great deal	*A lot*	*A moderate amount*	*A little*	*None at all*

12. How much did the role-play simulation help you better understand the viewpoints of different stakeholders in dealing with climate change in Cambridge?

1 -	2 -	3 -	4 -	5
A great deal	*A lot*	*A moderate amount*	*A little*	*None at all*

13. How much did the role-play simulation improve your ability to imagine how stakeholders in Cambridge could reach agreement on how to proceed with climate adaptation?

1 -	2 -	3 -	4 -	5
A great deal	*A lot*	*A moderate amount*	*A little*	*None at all*

14. What is your gender?
    a. Male
    b. Female
    c. Other: ______________

15. What is your age group?
    a. 19 or under
    b. 20–29
    c. 30–39
    d. 40–49
    e. 50–59
    f. 60+

16. How would you describe your political viewpoint?
    a. Extremely conservative
    b. Conservative
    c. Slightly conservative
    d. Moderate, middle of the road
    e. Slightly liberal
    f. Liberal
    g. Extremely liberal
    h. Other: ________________

17. Do you belong to any non-profit groups that regularly advocate on behalf of environmental conservation and/or protection? (Can select multiple options.)
    a. No
    b. Yes, a national group (like The Nature Conservancy or National Audubon Society, etc.)
    c. Yes, a local group (like a watershed alliance or local conservation committee, etc.)
    d. Yes, other: ______________________

18. What is the highest level of education you have completed?
    a. No formal schooling completed
    b. High school graduate (or equivalent)
    c. Some college or associate's degree (AA, etc.)
    d. Bachelor's degree (BA, BS, AB, etc.)
    e. Master's degree (MA, MSc)
    f. Professional or doctoral degree (MD, JD, PhD, etc.)
    g. Other: ________________

19. Could you estimate your household's total income for 2016? (Optional)
    a. Less than $14,999
    b. $15,000 to $34,999
    c. $25,000 to $34,999
    d. $35,000 to $49,999
    e. $50,000 to $74,999
    f. $75,000 to $99,999
    g. $100,000 to $149,999
    h. $150,000 or more

Appendix E

ROLE-PLAY SIMULATION FACILITATOR QUESTIONNAIRE

Name ____________________ **Date** ___________ **Table Number** _________

1. Did your table reach agreement? (YES / NO)

- If there was an agreement, was it unanimous? (YES / NO)
  - If not, which stakeholder did not take part in the agreement?
  - Which strategies were included or emphasized in the final agreement?
- If there wasn't an agreement, what (or who) was the obstacle to getting agreement?

2. Can you briefly describe what carried weight in terms of <u>process</u> or <u>substance</u> at the table you facilitated in reaching the final outcome?

Process explanation examples:

The people at my table were very cooperative.

They listened to each other and respected each other's views.

One of the parties offered a very persuasive argument and the rest of us went along.

There was only one person who didn't agree with rest of us.

The people who spoke seemed very sure of what they were saying.

Substantive explanation examples:

It wasn't hard to narrow down the list to the two most important strategies.

By tweaking one (or two) of the strategies, we were able to get everyone on board.

Most of us were clear on which strategies offered the greatest benefits to the whole city (or which were too costly relative to the benefits).

To get political agreement we had to include one of the strategies, even though one of the others might actually produce a better result.

The technical material made clear which would be the best strategies to include.

3. Did anything else occur at your table that you found interesting?

Appendix F

ROLE-PLAY SIMULATION DEBRIEFING SCRIPT

1. Before the game, had you thought about the public health impacts of climate change?
   Has your thinking changed? How?
2. Given the discussion during the game, which climate-related health impacts are you most worried about affecting you, your family, or your community? Why?
3. Most people haven't been thinking about climate change in terms of the increased risks to human health that flooding and heat waves might cause. Do you think talking about the city's vulnerability to climate change in terms of public health risks is a good idea? Why?
4. In the game, we were all talking about which strategies would be the best for the city of Mapleton to choose and why. Which of the five strategies do you think Cambridge should prioritize? Why?
5. It will not be easy to reduce the city's vulnerability to climate risks or enhance the city's resilience. There are things that might best be handled by the public sector (government) and other risk management efforts that should be the responsibility of citizens and property owners. What, in your view, are the government's responsibilities for trying to reduce the risks of climate change and what are the responsibilities of citizens and private property owners? Why?

Appendix G

ROLE-PLAY SIMULATION POST-GAME INTERVIEW QUESTIONS

1. Tell me a little bit about who you are and why you attended the [date and location] workshop.
   a. What is your name, title, and organization?
   b. What were you hoping to get from workshop?

2. What do you remember from playing the role-play simulation?
   a. Please tell us everything you remember from the game and workshop.
   b. Did you find the role-play simulation enjoyable? Interesting? What were the most striking elements?

(optional)

   c. What role did you play? Was the role you played and related interests similar or quite different from your real-life role and interests?

[Provide refresher on] date of game, outcome of game, table results

3. Prior to the workshop, what thoughts did you have about how climate change might affect Cambridge?
   a. How would you describe your thinking/concern about climate change impacts on Cambridge before the workshop?
   b. Before the game, had you thought about the public health impacts of climate change?
   c. How would you describe your policy support for climate adaptation in Cambridge before the workshop?
      i) What climate risks should Cambridge prioritize? [If needed, examples: property damage; business disruption; increased health risks; infrastructure damage; ecosystem degradation; disproportionate impacts to vulnerable populations] Why?
      ii) What policy approaches should Cambridge take? [If needed, examples: information provision; emergency preparedness; resource allocation; rules and regulations; prevention] Why?

4. Did the role-play simulation make you think or feel differently in any way on what you would recommend that Cambridge do to plan for climate change? Why?
   a. concern about climate impacts
   b. concern about public health
   c. policy recommendations for adaptation in Cambridge

5. Most people have not considered climate change in terms of public health risks from increased flooding and extreme heat. Do you think talking about Cambridge's vulnerability to climate change in terms of public health risks is a good idea? Why?

6. Is there anything else that you took away from participating in the simulation? [Can ask about climate risks, engaging diverse stakeholders in planning for adaptation, health impacts of climate change, etc.]
   a. Please explain how the role-play simulation affected you in this way.
   b. Do you think that these lessons could easily be learned in other ways?

7. Do you have any other recommendations on how to best engage diverse populations in climate adaptation planning and policymaking processes?
   a. What resonates with people?

Extra Potential Questions

Since participating in the workshop, have you talked with anyone (family, friends, colleagues) about the experience?
a. What did you tell them?

Was the role you played and related interests similar or quite different from your real-life role and interests?
a. How did it feel to "step into someone else's shoes" for a while?

Did the simulation make you think or feel differently in any way about the local climate risks that Cambridge is trying to prepare for?

Given the discussion during the game, which climate-related health impacts are you most worried will affect you, your family, or your community? Why?

Appendix H

ROLE-PLAY SIMULATION WORKSHOP SAMPLE CHARACTERISTICS (N = 112)

Variables		RPS Workshop Sample	City[a] Statistics
Gender	Male	47.6%	48.4%
	Female	52.4%	51.6%
Age	19 or under	1.0%	18.1%
	20–29	13.5%	30.5%
	30–39	18.3%	18.1%
	40–49	15.4%	9.7%
	50–59	21.2%	8.4%
	60+	30.8%	15.3%
Political Viewpoint[b]	Conservative	11.5%	26.7%
	Moderate	7.7%	36.6%
	Liberal	80.8%	32.5%
Education	High school graduate (or equivalent)	2.9%	9.3%
	Some college or associate's degree (AA)	2.9%	9.9%
	Bachelor's degree (BA, BS, AB, etc.)	24.0%	29.5%
	Master's degree (MA, MSc)	70.2%	45.2%
	Professional or doctoral degree (MD, PhD, etc.)		
Industry	Agriculture, forestry, fishing and hunting, mining	0%	0.2%
	Construction	3.1%	0.9%
	Manufacturing	1.0%	5.4%
	Wholesale trade	0%	1.1%
	Retail trade	0%	5.4%
	Transportation, warehousing, and utilities	2.1%	1.6%
	Management and administrative services	3.1%	20.5%
	Information, finance, insurance, real estate	13.5%	9.9%
	Educational services and healthcare	36.5%	41.8%
	Arts, entertainment, recreation, hospitality	3.1%	6.4%
	Public administration	5.2%	3.2%
	Other	32.3%	3.7%
Household Income[c]	Less than $14,999	3.5%	13.0%
	$15,000 to $24,999	10.3%	7.2%
	$25,000 to $34,999	2.3%	6.7%
	$35,000 to $49,999	5.8%	8.6%

(continued)

Variables	RPS Workshop Sample	City[a] Statistics
$50,000 to $74,999	12.6%	14.0%
$75,000 to $99,999	12.6%	12.3%
$100,000 to $149,999	25.2%	16.6%
$150,000 or more	27.6%	21.6%

[a] 2010–2014 American Community Survey 5-Year Estimates, U.S. Census Bureau (2015).
[b] Statewide (MA) statistics from Gallup State of the States 2015 used as proxy for city statistics.
[c] Workshop sample n = 94 for optional question on household income; city statistics given in 2014 inflation-adjusted dollars.

Appendix I

DIGITAL GAME BEFORE-SURVEY QUESTIONS

Q1. How much do you know about global warming or climate change?

- ○ I have never heard of it.
- ○ I know something about it.
- ○ I know a great deal about it.
- ○ Not sure.

Q2. How serious of a threat is climate change to you?

- ○ 1: Not at all serious
- ○ 2: Not very serious
- ○ 3: Somewhat serious
- ○ 4: Very serious
- ○ 5: Extremely serious

Q3. How concerned are you about the possible impacts a changing climate might have on the city of Cambridge?

- ○ 1: Not at all concerned
- ○ 2: Not very concerned
- ○ 3: Somewhat concerned
- ○ 4: Very concerned
- ○ 5: Extremely concerned

Q4. How willing would you be to pay higher taxes so that Cambridge can reduce climate change risks?

- ○ 1: Not at all willing
- ○ 2: Not very willing
- ○ 3: Somewhat willing
- ○ 4: Very willing
- ○ 5: Extremely willing

Q5. The city of Cambridge is looking to reduce its climate risks. Which climate change impact do you think Cambridge should prioritize in its policies?

- ○ Individual property damage
- ○ Local business disruption
- ○ Increased risks of disease, hospitalization, and death
- ○ Disproportionate impacts to people less able to prepare for climate impacts
- ○ Water, energy, and transportation infrastructure damage
- ○ Watershed, forest, and ecosystem degradation

Appendix J

DIGITAL GAME AFTER-SURVEY QUESTIONS

Q1. How much do you know about global warming or climate change?

- ○ I have never heard of it.
- ○ I know something about it.
- ○ I know a great deal about it.
- ○ Not sure.

Q2. How serious of a threat is climate change to you?

- ○ 1: Not at all serious
- ○ 2: Not very serious
- ○ 3: Somewhat serious
- ○ 4: Very serious
- ○ 5: Extremely serious

Q3. How concerned are you about the possible impacts a changing climate might have on the city of Cambridge?

- ○ 1: Not at all concerned
- ○ 2: Not very concerned
- ○ 3: Somewhat concerned
- ○ 4: Very concerned
- ○ 5: Extremely concerned

Q4. How willing would you be to pay higher taxes so that Cambridge can reduce climate change risks?

- ○ 1: Not at all willing
- ○ 2: Not very willing
- ○ 3: Somewhat willing
- ○ 4: Very willing
- ○ 5: Extremely willing

Q5. The city of Cambridge is looking to reduce its climate risks. Which climate change impact do you think Cambridge should prioritize in its policies?

- ○ Individual property damage
- ○ Local business disruption
- ○ Increased risks of disease, hospitalization, and death
- ○ Disproportionate impacts to people less able to prepare for climate impacts
- ○ Water, energy, and transportation infrastructure damage
- ○ Watershed, forest, and ecosystem degradation

Q6. How much did the digital game help you learn about the health impacts of climate change?

- ○ None at all
- ○ A little
- ○ A moderate amount
- ○ A lot
- ○ A great deal

Q7. How much did the digital game help you better understand the viewpoints of different stakeholders in dealing with climate change in Cambridge?

- ○ None at all
- ○ A little
- ○ A moderate amount
- ○ A lot
- ○ A great deal

Q8. How much did the digital game improve your ability to imagine how stakeholders in Cambridge could reach agreement on how to proceed with climate adaptation?

- ○ None at all
- ○ A little
- ○ A moderate amount
- ○ A lot
- ○ A great deal

Q9. What is your gender?

- ○ Male
- ○ Female
- ○ Other: ______________________________

Q10. What is your age?

- ○ 19 or under
- ○ 20–29
- ○ 30–39
- ○ 40–49
- ○ 50–59
- ○ 60+

Q11. How would your describe your political viewpoint?

- ○ Extremely conservative
- ○ Conservative
- ○ Slightly conservative
- ○ Moderate, middle of the road
- ○ Slightly liberal
- ○ Liberal
- ○ Extremely liberal
- ○ Other: ______________________________

Q12. Do you belong to any non-profit groups that regularly advocate on behalf of environmental conservation and/or protection? (Please select all that apply.)

- □ No
- □ Yes, a national group (like The Nature Conservatory or National Aubudon Society, etc.)
- □ Yes, a local group (like a watershed alliance or local conservation committee, etc.)
- □ Yes, other: ______________________________

Q13. What industry most accurately describes your primary field?

- ○ Agriculture, forestry, fishing and hunting, mining
- ○ Construction
- ○ Manufacturing
- ○ Wholesale trade
- ○ Retail trade
- ○ Transportation, warehousing, and utilities
- ○ Finance, insurance
- ○ Real estate
- ○ Management and administrative services
- ○ Education
- ○ Healthcare and social assistance
- ○ Arts, entertainment, recreation, hospitality
- ○ Public administration
- ○ Other: ______________________________

Q14. What is the highest level of education you have completed?

- ○ No formal schooling completed
- ○ High school graduate (or equivalent)
- ○ Some college or associate's degree (AA, etc.)
- ○ Bachelor's degree (BA, BS, AB, etc.)
- ○ Master's degree (MA, MSc)
- ○ Professional or doctoral degree (MD, JD, PhD, etc.)
- ○ Other: ______________________________

Appendix K

DIGITAL GAME POST-GAME INTERVIEW QUESTIONS

1. Tell me a little bit about who you are and why you played the digital game.
   a. What is your name, title, and organization?
   b. What were you hoping to get from the digital game?
2. What do you remember from playing the digital game?
   a. Please tell us everything you remember from the digital game.
   b. Did you find the digital game enjoyable? Interesting? What were the most striking elements?
3. Prior to playing the digital game, what thoughts did you have about how climate change might affect Cambridge?
   a. How would you describe your thinking/concern about climate change impacts on Cambridge before the digital game?
   b. Before the game, had you thought about the public health impacts of climate change?
   c. How would you describe your policy support for climate adaptation in Cambridge before the workshop?
      i) What climate risks should Cambridge prioritize? [If needed, examples: property damage; business disruption; increased health risks; infrastructure damage; ecosystem degradation; disproportionate impacts to vulnerable populations] Why?
      ii) What policy approaches should Cambridge take? [If needed, examples: information provision; emergency preparedness; resource allocation; rules and regulations; prevention] Why?
4. Did the digital game make you think or feel differently in any way on what you would recommend that Cambridge do to plan for climate change? Why?
   a. concern about climate impacts
   b. concern about public health
   c. policy recommendations for adaptation in Cambridge
5. Most people have not considered climate change in terms of public health risks from increased flooding and extreme heat. Do you think talking about Cambridge's vulnerability to climate change in terms of public health risks is a good idea? Why?

6. Is there anything else that you took away from playing the digital game? [Can ask about climate risks, engaging diverse stakeholders in planning for adaptation, health impacts of climate change, etc.]
   a. Please explain how the digital game affected you in this way.
   b. Do you think that these lessons could easily be learned in other ways?

7. Do you have any other recommendations on how to best engage diverse populations in climate adaptation planning?
   a. What resonates with people?

Extra Potential Questions

Since playing the digital game, have you talked with anyone (family, friends, colleagues) about the experience?

a. What did you tell them?

Did the digital game make you think or feel differently in any way about the local climate risks that Cambridge is trying to prepare for?

Given what you saw in the digital game, which climate-related health impacts are you most worried will affect you, your family, or your community? Why?

Appendix L

DIGITAL GAME SAMPLE CHARACTERISTICS (N = 105)

Variables		Digital Game Sample	City[a] Statistics
Gender	Male	28.6%	48.4%
	Female	70.5%	51.6%
Age	29 or under	7.7%	48.6%
	30–39	14.3%	18.1%
	40–49	19.1%	9.7%
	50–59	19.1%	8.4%
	60+	40.0%	15.3%
Political Viewpoint[b]	Conservative	21.9%	26.7%
	Moderate	19.1%	36.6%
	Liberal	59.1%	32.5%
Education	High school graduate (or equivalent)	1.0%	9.3%
	Some college or associate's degree (AA)	2.9%	9.9%
	Bachelor's degree (BA, BS, AB, etc.)	26.0%	29.5%
	Master's degree (MA, MSc, MBA, etc.)	88.5%	45.2%
	Professional or doctoral degree (MD, PhD, etc.)		
Industry	Agriculture, forestry, fishing and hunting, mining	1.0%	0.2%
	Construction	1.9%	0.9%
	Manufacturing	0.0%	5.4%
	Wholesale trade	0.0%	1.1%
	Retail trade	1.0%	5.4%
	Transportation, warehousing, and utilities	10.0%	1.6%
	Management and administrative services	9.7%	20.5%
	Information, finance, insurance, real estate	8.7%	9.9%
	Educational services and healthcare	33.0%	41.8%
	Arts, entertainment, recreation, hospitality	7.8%	6.4%
	Public administration	8.7%	3.2%
	Other	28.2%	3.7%

[a] 2010–2014 American Community Survey 5-Year Estimates, U.S. Census Bureau (2015).

[b] Statewide (MA) statistics from Gallup State of the States 2015 used as proxy for city statistics.

REFERENCES

Abt, C. C. 1987. *Serious Games*. Lanham, MD: University Press of America.

Ajzen, I. 1989. "Attitude Structure and Behavior." In *Attitude Structure and Function*, edited by A. R. Pratkanis, S. J. Breckler, and A. G. Greenwald, 241–74. Hillsdale, NJ: Erlbaum.

Akerlof, K., DeBono, R., Berry, P., Leiserowitz, A., Roser-Renouf, C., Clarke, K.-L., Rogaeva, A., Nisbet, M. C., Weathers, M. R., and Maibach, E. W. 2010. "Public Perceptions of Climate Change as a Human Health Risk: Surveys of the United States, Canada and Malta." *International Journal of Environmental Research and Public Health* 7 (6): 2559–606.

Allport, G. W. 1964. "Mental Health: A Generic Attitude." *Journal of Religion and Health* 4 (1): 7–21.

Amadeo, K. 2017. "Hurricane Harvey Facts, Damage and Costs." *The Balance*, September 30. Retrieved from https://www.thebalance.com/hurricane-harvey-facts-damage-costs-4150087.

American Public Health Association. 2018. *Adaptation in Action Part II: Updated Grantee Success Stories from CDC's Climate and Health Program.* https://www.apha.org/-/media/files/pdf/topics/climate/adaptation_in_action_part_2.ashx?la=en&hash=87A791182153A590EE7C5C97AE94EEC2691EFD6E.

Anenberg, S. C., Schwartz, J., Shindell, D. T., Amann, M., Faluvegi, G. S., Klimont, Z., Janssens-Maenhout, G., Pozzoli, L., Van Digenen, R., Vignati, E., Emberson, L., Muller, N. Z., West, J. J., Williams, M., Demkine, V., Hicks, W. K., Kuylenstierna, J., Raes, F., and Ramanathan, V. 2012. "Global Air Quality and Health Co-benefits of Mitigating Near-Term Climate Change through Methane and Black Carbon Emission Controls." *Environmental Health Perspectives* 120 (6): 831–39. https://doi.org/10.1289/ehp.1104301.

Anguelovski, I., and Carmin, J. 2011. "Something Borrowed, Everything New: Innovation and Institutionalization in Urban Climate Governance." *Current Opinion in Environmental Sustainability* 3 (3): 169–75.

Arcaya, M. C., Arcaya, A. L., and Subramanian, S. V. 2015. "Inequalities in Health: Definitions, Concepts, and Theories." *Global Health Action* 8 (1). https://doi.org/10.3402/gha.v8.27106.

Argyris, C., and Schon, D. 1978. *Organizational Learning: A Theory of Action Approach.* Reading, MA: Addison-Wesley.

Arnstein, S. R. 1969. "A Ladder of Citizen Participation." *Journal of the American Institute of Planners* 35 (4): 216–24.

Aslin, H., and Brown, V. 2004. *Towards Whole of Community Engagement: A Practical Toolkit.* Canberra: Murray-Darling Basin Commission. Retrieved from https://www.mdba.gov.au/sites/default/files/archived/mdbc-S-E-reports/1831_towards_whole_of_community_engagement_toolkit.pdf.

Astor, M. 2017. "'Flesh-Eating Bacteria' From Harvey's Floodwaters Kill a Woman." *New York Times*, September 28. Retrieved from https://www.nytimes.com/2017/09/28/health/necrotizing-fasciitis-houston-texas.html.

Bagozzi, R. P., Tybout, A. M., Craig, C. S., and Sternthal, B. 1979. "The Construct Validity of the Tripartite Classification of Attitudes." *Journal of Marketing Research* 16 (1): 88–95.

Balbus, J. M., and Malina, C. 2009. "Identifying Vulnerable Subpopulations for Climate Change Health Effects in the United States." *Journal of Occupational and Environmental Medicine* 51 (1): 33–37.

Bandura, A. 1998. "Personal and Collective Efficacy in Human Adaptation and Change." In *Advances in Psychological Science*, vol. 1, edited by J. G. Adair, D. Belanger, and K. L. Dion, 51–71. Hove: Psychology Press.

Beierle, T. C. 1999. "Using Social Goals to Evaluate Public Participation in Environmental Decisions." *Review of Policy Research* 16 (3–4): 75–103. https://doi.org/10.1111/j.1541-1338.1999.tb00879.x.

Bellotti, F., Kapralos, B., Lee, K., Moreno-Ger, P., and Berta, R. 2013. "Assessment in and of Serious Games: An Overview." *Advances in Human-Computer Interaction* 2013: 1–11.

Benhabib, S. 1996. *Democracy and Difference: Contesting the Boundaries of the Political.* Princeton, NJ: Princeton University Press.

Berkeley, E. P. 1968. "The New Gamesmanship." *Architectural Forum* 58: 58–63.

Best, S. J., and Krueger, B. S. 2005. "Analyzing the Representativeness of Internet Political Participation." *Political Behavior* 27 (2): 183–216.

Betsill, M., and Bulkeley, H. 2007. "Looking Back and Thinking Ahead: A Decade of Cities and Climate Change Research." *Local Environment* 12 (5): 447–56. https://doi.org/10.1080/13549830701659683.

Bierbaum, R., Smith, J. B., Lee, A., Blair, M., Carter, L., Chapin, F. S., III, Fleming, P., Ruffo, S., Stults, M., McNeeley, S., Wasley, E., and Verduzco, L. 2013. "A Comprehensive Review of Climate Adaptation in the United States: More than Before, but Less than Needed." *Mitigation and Adaptation Strategies for Global Change* 18 (3): 361–406.

Biggs, S. D. 1989. *Resource-Poor Farmer Participation in Research: A Synthesis of Experiences from Nine National Agricultural Research Systems.* The Hague, Netherlands: International Service for National Agricultural Research.

Bloomfield, D., Collins, K., Fry, C., and Munton, R. 2001. "Deliberation and Inclusion: Vehicles for Increasing Trust in UK Public Governance?" *Environment and Planning C: Government and Policy* 19 (4): 501–13.

Bollen, J., Guay, B., Jamet, S., and Corfee-Morlot, J. 2009. "Co-benefits of Climate Change Mitigation Policies: Literature Review and New Results." OECD Economics Department Working Paper No. 693. Retrieved from https://ideas.repec.org/p/oec/ecoaaa/693-en.html.

Boneau, C. A. 1960. "The Effects of Violations of Assumptions Underlying the T Test." *Psychological Bulletin* 57 (1): 49–64.

Breckler, S. J. 1984. "Empirical Validation of Affect, Behavior, and Cognition as Distinct Components of Attitude." *Journal of Personality and Social Psychology* 47 (6): 1191–205.

Brown, H. L., Proust, K., Spickett, J., and Capon, A. 2011. "The Potential Role of Health Impact Assessment in Tackling the Complexity of Climate Change Adaptation for Health." *Health Promotion Journal of Australia* 22 (4): 48–53.

Brunner, R., and Nordgren, J. 2016. "Climate Adaptation as an Evolutionary Process: A White Paper." In *Living with Climate Change: How Communities Are Surviving and Thriving in a Changing Climate*, edited by J. A. Bullock, G. D. Haddow, K. S. Haddow, and D. P. Coppola, 134–44. Boca Raton, FL: CRC Press.

Bulkeley, H., and Betsill, M. M. 2005. "Rethinking Sustainable Cities: Multilevel Governance and the 'Urban' Politics of Climate Change." *Environmental Politics* 14 (1): 42–63. https://doi.org/10.1080/0964401042000310178.

California Department of Public Health (CDPH). 2012. *Climate Action for Health: Integrating Public Health into Climate Action Planning.* California Department of Public Health. Retrieved from https://www.cdph.ca.gov/Programs/OHE/CDPH%20Document%20Library/CCHEP-General/CDPH-2012-Climate-Action-for-Health_accessible.pdf.

Carmin, J., Nadkarni, N., and Rhie, C. 2012. *Progress and Challenges in Urban Climate Adaptation Planning: Results of a Global Survey.* Cambridge, MA: Massachusetts Institute of Technology.

Carson, L. 2008. "The IAP2 Spectrum: Larry Susskind in Conversation with IAP2 Members." *International Journal of Public Participation* 2 (2): 67–84.

Cash, D., Clark, W. C., Alcock, F., Dickson, N. M., Eckley, N., and Jäger, J. 2002. "Salience, Credibility, Legitimacy and Boundaries: Linking Research, Assessment and Decision Making." *KSG Working Paper Series RWP02–046*. Retrieved from http://papers.ssrn.com/sol3/papers.cfm?abstract_id=372280.

Chan, M. 2007. "Climate Change and Health: Preparing for Unprecedented Challenges." The 2007 David E. Barmes Global Health Lecture, Bethesda, Maryland, USA, December 10, 2007. World Health Organization, Director-General Speeches. Retrieved from https://videocast.nih.gov/Summary.asp?File=14197&bhcp=1.

Chong, D. 1993. "How People Think, Reason, and Feel about Rights and Liberties." *American Journal of Political Science* 37 (3): 867–99. https://doi.org/10.2307/2111577.

City of Cambridge. 2002. *Climate Protection Plan: Local Actions to Reduce Greenhouse Gas Emissions*. Retrieved from http://www.cambridgema.gov/~/media/Files/CDD/Climate/climateplans/climate_plan.ashx.

———. 2014. *The City of Cambridge: Community Health Assessment*. Cambridge Public Health Department. Retrieved from http://www.cambridgepublichealth.org/publications/FinalCambridgeCHAreport.pdf.

———. 2015a. *Climate Change Vulnerability Assessment (Part I)*. Retrieved from http://www.cambridgema.gov/CDD/Projects/Climate/~/media/307B044E0EC5492BB92B2D8FA003ED25.ashx.

———. 2015b. *The Getting to Net Zero Framework*. Retrieved from http://www.cambridgema.gov/cdd/projects/climate/~/media/D74193AF8DAC4A57AC96E2A53946B96B.ashx.

———. 2015c. *Vulnerable Population Ranking Memorandum*. Retrieved from http://www.cambridgema.gov/CDD/Projects/Climate/~/media/8FA4C85FCCFA4043B779729B05999361.ashx.

———. 2017. *Climate Change Vulnerability Assessment (Part II)*. Retrieved from https://www.cambridgema.gov/~/media/Files/CDD/Climate/vulnerabilityassessment/Final-Report_CCVA-Part-2_MAR-2017_FINAL-2_web.pdf.

Cohen, J. 1988. *Statistical Power Analysis for the Behavioral Sciences*, 2nd ed. Hillsdale, NJ: Erlbaum.

Connolly, T. M., Boyle, E. A., MacArthur, E., Hainey, T., and Boyle, J. M. 2012. "A Systematic Literature Review of Empirical Evidence on Computer Games and Serious Games." *Computers & Education* 59 (2): 661–86.

Cornwall, A. 2008. "Unpacking 'Participation': Models, Meanings and Practices." *Community Development Journal* 43 (3): 269–83.

County of San Luis Obispo. 2011. *Climate Action Plan*. San Luis Obispo, CA: County of San Luis Obispo.

Creswell, J. W., and Clark, V. L. P. 2007. *Designing and Conducting Mixed Methods Research*. Thousand Oaks, CA: SAGE.

Day, D. 1997. "Citizen Participation in the Planning Process: An Essentially Contested Concept?" *CPL Bibliography* 11 (3): 421–34.

de Souza Briggs, X. 2008. *Democracy as Problem Solving: Civic Capacity in Communities across the Globe*. Cambridge, MA: MIT Press.

de Suarez, J. M., Suarez, P., Bachofen, C., Fortugno, N., Goentzel, J., Gonçalves, P., Grist, N., Macklin, C., de Suarez, J. M., Pfeifer, K., Schweizer, S., Suarez, P., van Aalst, M., and Virji, H. 2012. *Games for a New Climate: Experiencing the Complexity of Future Risks*. Pardee Center Task Force Report. Retrieved from http://www.oars3rivers.org/sites/default/files/Games-for-a-New-Climate-TF-Nov2012s.pdf.

Dessai, S., and Hulme, M. 2004. "Does Climate Adaptation Policy Need Probabilities?" *Climate Policy* 4 (2): 107–28.

Devitt, M. 2019. "AAFP Joins U.S. Call to Action on Climate Change." *American Academy of Family Physicians News*, July 2. Retrieved from https://www.aafp.org/news/health-of-the-public/20190702climatecall2action.html.

Dolin, E. J., and Susskind, L. E. 1992. "A Role for Simulations in Public Policy Disputes: The Case of National Energy Policy." *Simulation & Gaming* 23 (1): 20–44.

Druckman, J. N. 2001a. "On the Limits of Framing Effects: Who Can Frame?" *Journal of Politics* 63 (4): 1041–66.

———. 2001b. "The Implications of Framing Effects for Citizen Competence." *Political Behavior* 23 (3): 225–56.

Druckman, J. N., and Nelson, K. R. 2003. "Framing and Deliberation: How Citizens' Conversations Limit Elite Influence." *American Journal of Political Science* 47 (4): 729–45.

Duane, T. P. 1999. *Shaping the Sierra: Nature, Culture, and Conflict in the Changing West.* Berkeley: University of California Press.

Dunlap, R. E., McCright, A. M., and Yarosh, J. H. 2016. "The Political Divide on Climate Change: Partisan Polarization Widens in the US. Environment." *Science and Policy for Sustainable Development* 58 (5): 4–23.

Dupuis, J., and Knoepfel, P. 2013. "The Adaptation Policy Paradox: The Implementation Deficit of Policies Framed as Climate Change Adaptation." *Ecology and Society* 18 (4): 31–46.

Eagly, A. H., and Chaiken, S. 2007. "The Advantages of an Inclusive Definition of Attitude." *Social Cognition* 25 (5): 582–602.

Ebi, K. L. 2013. "Health in the New Scenarios for Climate Change Research." *International Journal of Environmental Research and Public Health* 11 (1): 30–46. https://doi.org/10.3390/ijerph110100030.

Emanuel, K. 2017. "Why It's Time to Stop Calling These Hurricane Disasters 'Natural.'" *Washington Post*, September 19. Retrieved from https://www.washingtonpost.com/news/energy-environment/wp/2017/09/19/why-its-time-to-stop-calling-these-hurricane-disasters-natural/?tid=ss_tw-bottom&utm_term=.5a5cdcdaa5e7.

Entertainment Software Association (ESA). 2017. *Essential Facts about the Computer and Video Game Industry*. Retrieved from https://www.theesa.com/esa-research/2017-essential-facts-about-the-computer-and-video-game-industry/.

Entman, R. M. 1993. "Framing: Toward Clarification of a Fractured Paradigm." *Journal of Communication* 43 (4): 51–58.

Environmental Protection Agency (EPA). 2016. How Will Climate Change Affect My Health? Retrieved from https://www3.epa.gov/climatechange/impacts/health/factsheets/.

Erhel, S., and Jamet, E. 2013. "Digital Game-Based Learning: Impact of Instructions and Feedback on Motivation and Learning Effectiveness." *Computers & Education* 67: 156–67.

Evans-Cowley, J., and Hollander, J. 2010. "The New Generation of Public Participation: Internet-based Participation Tools." *Planning Practice & Research* 25 (3): 397–408. https://doi.org/10.1080/02697459.2010.503432.

Farrington, J., and Bebbington, A. 1993. *Reluctant Partners?: Non-Governmental Organizations, the State and Sustainable Agricultural Development.* New York: Routledge.

Fearon, J. D. 1998. "Deliberation as Discussion." In *Deliberative Democracy*, edited by J. Elster, 44–68, 44, 56. Cambridge: Cambridge University Press.

Few, R., Brown, K., and Tompkins, E. L. 2007. "Public Participation and Climate Change Adaptation: Avoiding the Illusion of Inclusion." *Climate Policy* 7 (1): 46–59.

Fiorina, M. P. 1999. "Extreme Voices: A Dark Side of Civic Engagement." In *Civic Engagement in American Democracy*, edited by T. Skocpol and M. P. Fiorina, 395, 405–13. Washington, DC: Brookings Institution Press.

Fischer, A., and Young, J. C. 2007. "Understanding Mental Constructs of Biodiversity: Implications for Biodiversity Management and Conservation." *Biological Conservation* 136 (2): 271–82.

Fischhoff, B. 2007. "Nonpersuasive Communication about Matters of Greatest Urgency: Climate Change." *Environmental Science & Technology* 41 (21): 7204–8.

Frumkin, H., Hess, J., Luber, G., Malilay, J., and McGeehin, M. 2008. "Climate Change: The Public Health Response." *American Journal of Public Health* 98 (3): 435–45.

Fung, A., and Wright, E. O. 2003. *Deepening Democracy: Institutional Innovations in Empowered Participatory Governance*, vol. 4. London: Verso.

Füssel, H. M. 2007. "Adaptation Planning for Climate Change: Concepts, Assessment Approaches, and Key Lessons." *Sustainability Science* 2 (2): 265–75.

Gallup. 2015. *State of the States.* Retrieved from http://news.gallup.com/poll/125066/state-states.aspx.

Gee, J. P. 2003. "What Video Games Have to Teach Us about Learning and Literacy." *Computers in Entertainment (CIE)* 1 (1): 20–20.

Georgetown Climate Center. 2016. "State and Local Adaptation Plans." Accessed April 18, 2017. https://www.georgetownclimate.org/adaptation/plans.html.

Gifford, R., and Comeau, L. A. 2011. "Message Framing Influences Perceived Climate Change Competence, Engagement, and Behavioral Intentions." *Global Environmental Change* 21 (4): 1301–7.

Gordon, E., and Baldwin-Philippi, J. 2014. "Playful Civic Learning: Enabling Reflection and Lateral Trust in Game-Based Public Participation." *International Journal of Communication* 8: 759–86.

Gosen, J., and Washbush, J. 2004. "A Review of Scholarship on Assessing Experiential Learning Effectiveness." *Simulation & Gaming* 35 (2): 270–93.

Graesser, A., Chipman, P., Leeming, F., and Biedenbach, S. 2009. "Deep Learning and Emotion in Serious Games." In *Serious Games: Mechanisms and Effects*, edited by U. Ritterfeld, M. Cody, and P. Vorderer, 81–100. New York: Routledge.

Grigg, N. 2017. "Flooding from Hurricane Harvey Causes a Host of Public Hhealth Concerns." *The Conversation*, August 30. Retrieved from https://theconversation.com/flooding-from-hurricane-harvey-causes-a-host-of-public-health-concerns-83134.

Hansen, J., Kharecha, P., Sato, M., Masson-Delmotte, V., Ackerman, F., Beerling, D. J., Hearty, P. J., Hoegh-Guldberg, O., Hsu, S.-L., Parmesan, C., Rockstrom, J., Rohling, E. J., Sachs, J., Smith, P., Steffen, K., Van Susteren, L., von Schuckmann, K., and Zachos, J. C. 2013. "Assessing 'Dangerous Climate Change': Required Reduction of Carbon Emissions to Protect Young People, Future Generations and Nature." *PloS One* 8 (12): e81648.

Harlan, S. L., and Ruddell, D. M. 2011. "Climate Change and Health in Cities: Impacts of Heat and Air Pollution and Potential Co-benefits from Mitigation and Adaptation." *Current Opinion in Environmental Sustainability* 3 (3): 126–34. https://doi.org/10.1016/j.cosust.2011.01.001.

Harries, T., and Penning-Rowsell, E. 2011. "Victim Pressure, Institutional Inertia and Climate Change Adaptation: The Case of Flood Risk." *Global Environmental Change* 21 (1): 188–97.

Hays, Robert T. "The Effectiveness of Instructional Games: A Literature Review and Discussion." No. NAWCTSD-TR-2005-004. Naval Air Warfare Center Training Systems Division, Orlando, Florida, 2005.

Hernandez, A., Zezima, K., and Achenbach, J. 2017. "Texas Faces Environmental Concerns as Wastewater, Drinking Water Systems Compromised." *Washington Post*, September 3. Retrieved from https://www.washingtonpost.com/news/post-nation/wp/2017/09/03/trump-administration-wants-to-tie-harvey-recovery-aid-to-debt-ceiling-legislation/?utm_term=.eb14637c526e.

Hilgard, E. R. 1980. "The Trilogy of Mind: Cognition, Affection, and Conation." *Journal of the History of the Behavioral Sciences* 16 (2): 107–117.

Hodgkinson, G. P., Maule, A. J., Bown, N. J., Pearman, A. D., and Glaister, K. W. 2002. "Further Reflections on the Elimination of Framing Bias in Strategic Decision Making." *Strategic Management Journal* 23 (11): 1069–76.

Hofstede, J. 2008. "Climate Change and Coastal Adaptation Strategies: The Schleswig-Holstein Perspective." *Baltica* 21 (1–2): 71–78.

Hornsey, M. J., Harris, E. A., Bain, P. G., and Fielding, K. S. 2016. "Meta-Analyses of the Determinants and Outcomes of Belief in Climate Change." *Nature Climate Change* 6: 622–26. https://doi.org/10.1038/nclimate2943.

Howe, P. D., Mildenberger, M., Marlon, J. R., and Leiserowitz, A. 2015. "Geographic Variation in Opinions on Climate Change at State and Local Scales in the USA." *Nature Climate Change* 5 (6): 596–603. https://doi.org/10.1038/nclimate2583.

Huang, W.-H. 2011. "Evaluating Learners' Motivational and Cognitive Processing in an Online Game-Based Learning Environment." *Computers in Human Behavior* 27 (2): 694–704.

Hughes, S. 2015. "A Meta-Analysis of Urban Climate Change Adaptation Planning in the U.S." *Urban Climate* 14 (Part 1): 17–29.

Hulme, M., Adger, W. N., Dessai, S., Lorenzoni, I., Naess, L., and Wreford, A. 2007. "Limits and Barriers to Adaptation: Four Propositions." Tyndall Center for Climate Change Research. Norwich: University of East Anglia.

Innes, J. E., and Booher, D. E. 2004. "Reframing Public Participation: Strategies for the 21st Century." *Planning Theory & Practice* 5 (4): 419–36.

IPCC. 2000. *Special Report on Emissions Scenarios.*

———. 2001. *Climate Change 2001: Impacts, Adaptation and Vulnerability.* Contribution of Working Group II to the Third Assessment Report of the Intergovernmental Panel on Climate Change, edited by J. J. McCarthy, O. F. Canziani, N. A. Leary, D. J. Dokken, and K. S. White.

———. 2012. *Managing the Risks of Extreme Events and Disasters to Advance Climate Change Adaptation: Special Report of the Intergovernmental Panel on Climate Change.*

———. 2014. *Climate Change 2014: Impacts, Adaptation, and Vulnerability (Vol. 1).* Retrieved from https://www.ipcc.ch/report/ar5/wg2/.

Jamieson, S. 2004. "Likert Scales: How to (Ab)use Them." *Medical Education* 38 (12): 1217–18.

Kahan, D. M., Peters, E., Dawson, E. C., and Slovic, P. 2013. "Motivated Numeracy and Enlightened Self-Government." *Behavioral Public Policy* 1: 54–86. Yale Law School, Public Law Working Paper No. 307. Retrieved from http://papers.ssrn.com/sol3/papers.cfm?abstract_id=2319992.

Kahan, D. M., Peters, E., Wittlin, M., Slovic, P., Ouellette, L. L., Braman, D., and Mandel, G. 2012. "The Polarizing Impact of Science Literacy and Numeracy on Perceived Climate Change Risks." *Nature Climate Change* 2 (10): 732–35.

Kaplan, S., and Healy, J. 2017. "Houston's Floodwaters Are Tainted, Testing Shows." *New York Times*, September 11. Retrieved from https://www.nytimes.com/2017/09/11/health/houston-flood-contamination.html.

Kaufman, S., Elliott, M., and Shmueli, D. 2003. "Frames, Framing and Reframing." *Beyond Intractability*. https://www.beyondintractability.org/essay/framing.

Ke, F. 2009. "A Qualitative Meta-Analysis of Computer Games as Learning Tools." In *Handbook of Research on Effective Electronic Gaming in Education*, vol. 1, edited by R. E. Ferdig, 1–32. Hershey, PA: Information Science Reference.

Kelly, M., Ferranto, S., Lei, S., Ueda, K., and Huntsinger, L. 2012. "Expanding the Table: The Web as a Tool for Participatory Adaptive Management in California Forests." *Journal of Environmental Management* 109: 1–11.

Kim, E. 2019. "Staten Island's $615 Million Sea Wall Project Secures Critical Funding." *Gothamist*, February 20. Retrieved from https://gothamist.com/2019/02/20/staten_islands_sea_wall.php.

Kim, T., Hong, H., and Magerko, B. 2010. "Designing for Persuasion: Toward Ambient Eco-Visualization for Awareness." In *Persuasive Technology*, edited by T. Ploug, P. Hasle, and H. Oinas-Kukkonen, 106–16. Berlin: Springer.

Kinney, P. L., Matte, T., Knowlton, K., Madrigano, J., Petkova, E., Weinberger, K., Quinn, A., Arend, M., and Pullen, J. 2015. "New York City Panel on Climate Change 2015 Report Chapter 5: Public Health Impacts and Resiliency." *Annals of the New York Academy of Sciences* 1336 (1): 67–88. https://doi.org/10.1111/nyas.12588.

Kinney, P., Little, E., Petkova, E., and Chen, K. 2015. *Public Health Assessment, Climate Change Vulnerability Assessment.* City of Cambridge. Retrieved from https://www.cambridgema.gov/CDD/Projects/Climate/~/media/190871A066004CEBA2198F83EDAF737E.ashx.

Kolb, D. A. 2014. *Experiential Learning: Experience as the Source of Learning and Development.* Upper Saddle River, NJ: FT Press.

Konisky, D. 2001. "Innovations in Public Participation and Environmental Decision Making: Examples from the Great Lakes Region." *Society &Natural Resources* 14 (9): 815–26.

Kothari, U. 2001. "Power, Knowledge and Social Control in Participatory Development." In *Participation: The New Tyranny?*, edited by B. Cooke and U. Kothari, 139–52. New York: Zed Books.

Kühberger, A. 1997. "Theoretical Conceptions of Framing Effects in Risky Decisions." In *Decision Making: Cognitive Models and Explanations*, edited by R. Ranyard, W. R. Crozier, and O. Svenson, 128–44. New York: Routledge.

Lane, K., Wheeler, K., Charles-Guzman, K., Ahmed, M., Blum, M., Gregory, K., Graber, M., Clark, N., and Matte, T. 2014. "Extreme Heat Awareness and Protective Behaviors in New York City." *Journal of Urban Health* 91 (3): 403–14.

Lee, J. J., Ceyhan, P., Jordan-Cooley, W., and Sung, W. 2013. "GREENIFY: A Real-World Action Game for Climate Change Education." *Simulation & Gaming* 44 (2–3): 349–65.

Lee, T. M., Markowitz, E. M., Howe, P. D., Ko, C.-Y., and Leiserowitz, A. A. 2015. "Predictors of Public Climate Change Awareness and Risk Perception around the World." *Nature Climate Change* 5 (11): 1014–20. https://doi.org/10.1038/nclimate2728.

Leiserowitz, A. 2006. "Climate Change Risk Perception and Policy Preferences: The Role of Affect, Imagery, and Values." *Climatic Change* 77 (1–2): 45–72.

Leiserowitz, A., and Smith, N. 2017. "Affective Imagery, Risk Perceptions, and Climate Change Communication." *Oxford Research Encyclopedia of Climate Science.* https://doi.org/10.1093/acrefore/9780190228620.013.307.

Leiserowitz, A., Maibach, E., Roser-Renouf, C., Feinberg, G., and Howe, P. D. 2013. *Global Warming's Six Americas in September 2012.* New Haven, CT: Yale Project on Climate Change Communication. Retrieved from https://environment.yale.edu/climate-communication-OFF/files/Six-Americas-September-2012.pdf.

Levin, I. P., and Gaeth, G. J. 1988. "How Consumers Are Affected by the Framing of Attribute Information before and after Consuming the Product." *Journal of Consumer Research* 15 (3): 374–78.

Levin, I. P., Schneider, S. L., and Gaeth, G. J. 1998. "All Frames Are Not Created Equal: A Typology and Critical Analysis of Framing Effects." *Organizational Behavior and Human Decision Processes* 76 (2): 149–88.

Ligeti, E., Wieditz, I., and Penney, J. 2007. Cities Preparing for *Climate Change: A Study of Six Urban Regions.* Toronto: Clean Air Partnership.

Light, J. 2008. "Taking Games Seriously." *Technology and Culture* 49 (2): 347–75.

Lowndes, V., Pratchett, L., and Stoker, G. 2001. "Trends in Public Participation: Part 1–Local Government Perspectives." *Public Administration* 79 (1): 205–22.

Luber, G., Knowlton, K., Balbus, J., Frumkin, H., Hayden, M., Hess, J., McGeehin, M., Sheats, N., Backer, L., Beard, C. B., Ebi, K. L., Maibach, E., Ostfeld, R. S., Wiedinmyer, C., Zielinski-Gutiérrez, E., and Ziska, L. 2014. "Human Health." In *Climate Change Impacts in the United States: The Third National Climate Assessment*, edited by J. M. Melillo, T. C. Richmond, and G. W. Yohe, 220–56. U.S. Global Research Program. Washington, DC: U.S. Government Printing Office.

Lyles, W., Berke, P., and Overstreet, K. H. 2018. "Where to Begin Municipal Climate Adaptation Planning? Evaluating Two Local Choices." *Journal of Environmental Planning and Management* 61 (11): 1994–2014.

Macnaghten, P., and Jacobs, M. 1997. "Public Identification with Sustainable Development: Investigating Cultural Barriers to Participation." *Global Environmental Change* 7 (1): 5–24.

Maibach, E. W., Kreslake, J. M., Roser-Renouf, C., Rosenthal, S., Feinberg, G., and Leiserowitz, A. A. 2015. "Do Americans Understand That Global Warming Is Harmful to Human Health? Evidence from a National Survey." *Annals of Global Health* 81 (3): 396–409. https://doi.org/10.1016/j.aogh.2015.08.010.

Maibach, E. W., Nisbet, M., Baldwin, P., Akerlof, K., and Diao, G. 2010. "Reframing Climate Change as a Public Health Issue: An Exploratory Study of Public Reactions." *BMC Public Health* 10, 299. https://doi.org/10.1186/1471-2458-10-299.

Mandarano, L. A. 2008. "Evaluating Collaborative Environmental Planning Outputs and Outcomes: Restoring and Protecting Habitat and the New York—New Jersey Harbor Estuary Program." *Journal of Planning Education and Research* 27 (4): 456–68.

Marinucci, G. D., Luber, G., Uejio, C. K., Saha, S., and Hess, J. J. 2014. "Building Resilience against Climate Effects—A Novel Framework to Facilitate Climate Readiness in Public Health Agencies." *International Journal of Environmental Research and Public Health* 11 (6): 6433–58.

Martin, A., and Sherington, J. 1997. "Participatory Research Methods—Implementation, Effectiveness and Institutional Context." *Agricultural Systems* 55 (2): 195–216.

Mayer, R. E., and Johnson, C. I. 2010. "Adding Instructional Features That Promote Learning in a Game-Like Environment." *Journal of Educational Computing Research* 42 (3): 241–65.

McElroy, T., and Seta, J. J. 2003. "Framing Effects: An Analytic–Holistic Perspective." *Journal of Experimental Social Psychology* 39 (6): 610–17.

McGonigal, J. 2011. *Reality Is Broken: Why Games Make Us Better and How They Can Change the World.* New York: Penguin.

McGuire, W. J. 1969. "The Nature of Attitudes and Attitude Change." In *The Handbook of Social Psychology*, 2nd ed., vol. 3, edited by G. Lindzey and E. Aronson, 136–314. Reading, MA: Addison-Wesley.

Mees, H. L., Driessen, P. P., and Runhaar, H. A. 2015. "'Cool' Governance of a 'Hot' Climate Issue: Public and Private Responsibilities for the Protection of Vulnerable Citizens against Extreme Heat." *Regional Environmental Change* 15 (6): 1065–79.

Melillo, J. M., Richmond, T. C., and Yohe, G. W. 2014. *Climate Change Impacts in the United States: The Third National Climate Assessment.* U.S. Global Change Research Program. Washington, DC: U.S. Government Printing Office. https://www.nrc.gov/docs/ML1412/ML14129A233.pdf.

Morton, T. A., Rabinovich, A., Marshall, D., and Bretschneider, P. 2011. "The Future That May (or May Not) Come: How Framing Changes Responses to Uncertainty in Climate Change Communications." *Global Environmental Change* 21 (1): 103–9.

Moser, S. C., and Dilling, L. 2004. "Making Climate Hot." *Environment: Science and Policy for Sustainable Development* 46 (10): 32–46.

Moser, S. C., and Ekstrom, J. A. 2010. "A Framework to Diagnose Barriers to Climate Change Adaptation." *Proceedings of the National Academy of Sciences of the United States of America* 107 (51): 22026–31.

Moser, S. C., and Pike, C. 2015. "Community Engagement on Adaptation: Meeting a Growing Capacity Need." *Urban Climate* 14 (Part 1): 111–15. https://doi.org/10.1016/j.uclim.2015.06.006.

Myers, T. A., Nisbet, M. C., Maibach, E. W., and Leiserowitz, A. A. 2012. "A Public Health Frame Arouses Hopeful Emotions about Climate Change." *Climatic Change* 113 (3–4): 1105–12. https://doi.org/10.1007/s10584-012-0513-6.

Nelson, N., and Wright, S. 1995. *Power and Participatory Development: Theory and Practice.* London: ITDG.

Nelson, T. E., Oxley, Z. M., and Clawson, R. A. 1997. "Toward a Psychology of Framing Effects." *Political Behavior* 19 (3): 221–46.

Nemet, G. F., Holloway, T., and Meier, P. 2010. "Implications of Incorporating Air-Quality Co-benefits into Climate Change Policymaking." *Environmental Research Letters* 5 (1). https://doi.org/10.1088/1748-9326/5/1/014007.

Nisbet, M. C. 2009. "Communicating Climate Change: Why Frames Matter for Public Engagement." *Environment: Science and Policy for Sustainable Development* 51 (2): 12–23.

Nisbet, M. C., and Scheufele, D. A. 2009. "What's Next for Science Communication? Promising Directions and Lingering Distractions." *American Journal of Botany* 96 (10): 1767–78.

Nordgren, J., Stults, M., and Meerow, S. 2016. "Supporting Local Climate Change Adaptation: Where We Are and Where We Need to Go." *Environmental Science & Policy* 66: 344–52.

Norman, G. 2010. "Likert Scales, Levels of Measurement and the 'Laws' of Statistics." *Advances in Health Sciences Education* 15 (5): 625–32.

O'Neill, M. S., Kinney, P. L., and Cohen, A. J. 2008. "Environmental Equity in Air Quality Management: Local and International Implications for Human Health and Climate Change." *Journal of Toxicology and Environmental Health A* 71 (9–10): 570–77.

Oreskes, N., and Conway, E. M. 2010. *Merchants of Doubt.* New York: Bloomsbury Press.

Patz, J. A., Gibbs, H. K., Foley, J. A., Rogers, J. V., and Smith, K. R. 2007. "Climate Change and Global Health: Quantifying a Growing Ethical Crisis." *EcoHealth* 4 (4): 397–405.

Patz, J., Campbell-Lendrum, D., Gibbs, H., and Woodruff, R. 2008. "Health Impact Assessment of Global Climate Change: Expanding on Comparative Risk Assessment Approaches for Policy Making." *Annual Review of Public Health* 29 (1): 27–39. https://doi.org/10.1146/annurev.publhealth.29.020907.090750.

Payne, H. 2016. "Engaging the Public in Climate Adaptation Planning: Lessons from Sixteen American Cities." MCP Thesis, Massachusetts Institute of Technology. Retrieved from https://dspace.mit.edu/handle/1721.1/105058.

Pearson, E. S. 1931. "The Analysis of Variance in Cases of Non-Normal Variation." *Biometrika* 23 (1–2): 114–33.

Pidgeon, N., and Fischhoff, B. 2011. "The Role of Social and Decision Sciences in Communicating Uncertain Climate Risks." *Nature Climate Change* 1 (1): 35–41.

Potter, R. B. 1985. *Urbanisation and Planning in the Third World: Spatial Perceptions and Public Participation.* London: Croom Helm.

Prensky, M. 2003. "Digital Game-Based Learning." *Computers in Entertainment (CIE)* 1 (1): 21–21.

Pretty, J. N. 1995. "Participatory Learning for Sustainable Agriculture." *World Development* 23 (8): 1247–63.

Randel, J. M., Morris, B. A., Wetzel, C. D., and Whitehill, B. V. 1992. "The Effectiveness of Games for Educational Purposes: A Review of Recent Research." *Simulation & Gaming* 23 (3): 261–76.

Reed, M. S. 2008. "Stakeholder Participation for Environmental Management: A Literature Review." *Biological Conservation* 141 (10): 2417–31.

Renn, O. 2006. "Participatory Processes for Designing Environmental Policies." *Land Use Policy* 23 (1): 34–43.

Rittel, H. W., and Webber, M. M. 1973. "Dilemmas in a General Theory of Planning." *Policy Sciences* 4 (2): 155–69.

Robinson, P. J. 2001. "On the Definition of a Heat Wave." *Journal of Applied Meteorology* 40 (4): 762–75.

Rosenberg, M. J., and Hovland, C. I. 1960. "Cognitive, Affective, and Behavioral Components of Attitudes." In *Attitude Organization and Change: An Analysis of Consistency among Attitude Components,* edited by M. J. Rosenberg, C. I. Hovland, W. J. McGuire, R. P. Abelson, and J. W. Brehm, 1–14. New Haven, CT: Yale University Press.

Rosenthal, J. K., and Brechwald, D. 2013. "Climate Adaptive Planning for Preventing Heat-related Health Impacts in New York City." In *Climate Change Governance,* edited by J. Knieling and W. L. Filho, 205–25. Berlin: Springer. https://doi.org/10.1007/978-3-642-29831-8_13.

Rosenzweig, C., and Solecki, W. 2010. "Introduction to Climate Change Adaptation in New York City: Building a Risk Management Response." *Annals of the New York Academy of Sciences* 1196 (1): 13–18. https://doi.org/10.1111/j.1749-6632.2009.05306.x.

Rosenzweig, C., Solecki, W. D., Blake, R., Bowman, M., Faris, C., Gornitz, V., Horton, R., Jacob, K., LeBlanc, A., Leichenko, R., Linkin, M., Major, D., O'Grady, M., Patrick, L., Sussman, E., Yohe, G., and Zimmerman, R. 2011. "Developing Coastal Adaptation to Climate Change in the New York City Infrastructure-Shed: Process, Approach, Tools, and Strategies." *Climatic Change* 106 (1): 93–127.

Rowe, G., and Frewer, L. J. 2000. "Public Participation Methods: A Framework for Evaluation." *Science, Technology, & Human Values* 25 (1): 3–29.

Rumbach, A. J., and Kudva, N. 2011. "Putting People at the Center of Climate Change Adaptation Plans: A Vulnerability Approach." *Risk, Hazards & Crisis in Public Policy* 2 (4): 1–23.

Rumore, D., Schenk, T., and Susskind, L. 2016. "Role-Play Simulations for Climate Change Adaptation Education and Engagement." *Nature Climate Change* 6 (8): 745–50.

Sailor, D., Shepherd, M., Sheridan, S., Stone, B., Kalkstein, L., Russell, A., Vargo, J., and Andersen, T. 2016. "Improving Heat-Related Health Outcomes in an Urban Environment with Science-Based Policy." *Sustainability* 8 (10): 1015. https://doi.org/10.3390/su8101015.

Sanchez, E., Cronick, K., and Wiesenfeld, E. 1988. "Psychosocial Variables and Participation: A Case Study." In *New Directions in Environmental Participation, Vol. 3*, edited by D. Canter, M. Kramping, and D. Stea, 1–17. Aldershot: Avebury.

Sawyer, B., and Smith, P. 2008. "Serious Games Taxonomy." Retrieved from http://www.dmill.com/presentations/serious-games-taxonomy-2008.pdf.

Schenk, T. E. W. 2015. "Institutionalizing Uncertainty: Exploring How Infrastructure Stakeholders Can Prepare for Uncertain Climate Futures." PhD dissertation at the Massachusetts Institute of Technology.

Schon, D. A., and Rein, M. 1995. *Frame Reflection: Toward the Resolution of Intractable Policy Controversies.* New York: Basic Books.

Shaffer, D. W. 2006. "Epistemic Frames for Epistemic Games." *Computers & Education* 46 (3): 223–34.

Shaw, A., Burch, S., Kristensen, F., Robinson, J., and Dale, A. 2014. "Accelerating the Sustainability Transition: Exploring Synergies between Adaptation and Mitigation in British Columbian Communities." *Global Environmental Change* 25: 41–51.

Shaw, C., Hales, S., Howden-Chapman, P., and Edwards, R. 2014. "Health Co-benefits of Climate Change Mitigation Policies in the Transport Sector." *Nature Climate Change* 4 (6): 427–33. https://doi.org/10.1038/nclimate2247.

Shi, L., Chu, E., and Debats, J. 2015. "Explaining Progress in Climate Adaptation Planning Across 156 U.S. Municipalities." *Journal of the American Planning Association* 81 (3): 191–202. https://doi.org/10.1080/01944363.2015.1074526.

Smit, B., and Wandel, J. 2006. "Adaptation, Adaptive Capacity and Vulnerability." *Global Environmental Change* 16 (3): 282–92.

Smith, S. M., and Levin, I. P. 1996. "Need for Cognition and Choice Framing Effects." *Journal of Behavioral Decision Making* 9 (4): 283–90.

Smith, T. W., P. Marsden, M. Hout, and J. Kim. 2017. *General Social Surveys, 1972–2016.* Chicago, IL: NORC at the University of Chicago.

Sniderman, P. M., and Theriault, S. M. 2004. "The Structure of Political Argument and the Logic of Issue Framing." In *Studies in Public Opinion: Attitudes, Nonattitudes, Measurement Error, and Change,* edited by W. E. Saris and P. M. Sniderman, 133–65. Princeton, NJ: Princeton University Press.

Solecki, W., Rosenzweig, C., Blake, R., de Sherbinin, A., Matte, T., Moshary, F., Rosenzweig, B., Arend, M., Gaffin, S., Bou-Zeid, E., Rule, K., Sweeney, G., and Dessy, W. 2015. "New York City Panel on Climate Change 2015 Report Chapter 6: Indicators and Monitoring." *Annals of the New York Academy of Sciences* 1336 (1): 89–106. https://doi.org/10.1111/nyas.12587.

Solomon, S., Plattner, G., Knutti, R., and Friedlingstein, P. 2009. "Irreversible Climate Change due to Carbon Dioxide Emissions." *Proceedings of the National Academy of Sciences of the United States of America* 106 (6): 1704–9.

Spence, Alexa, and Nick Pidgeon. 2010. "Framing and communicating climate change: The effects of distance and outcome frame manipulations." *Global Environmental Change* 20 (4): 656–67.

Squire, K. 2003. "Video Games in Education." *Int. J. Intell. Games & Simulation* 2 (1): 49–62.

———. 2006. "From Content to Context: Videogames as Designed Experience." *Educational Researcher* 35 (8): 19–29.

Sterman, J. D. 2011. "Communicating Climate Change Risks in a Skeptical World." *Climatic Change* 108 (4): 811–26. https://doi.org/10.1007/s10584-011-0189-3.

Sterman, J. D., and Sweeney, L. B. 2007. "Understanding Public Complacency about Climate Change: Adults' Mental Models of Climate Change Violate Conservation of Matter." *Climatic Change* 80 (3–4): 213–38.

Stone, B. 2012. *The City and the Coming Climate: Climate Change in the Places We Live*. New York: Cambridge University Press.

Sunstein, C. R. 2000. "Cognition and Cost-Benefit Analysis." *Journal of Legal Studies* 29 (S2): 1059–103.

Susskind, L. 1987. *Breaking the Impasse: Consensual Approaches to Resolving Public Disputes*. New York: Basic Books.

———. 2001. *Better Environmental Policy Studies: How to Design and Conduct More Effective Analysis*. Washington, DC: Island Press.

———. 2010. "Policy & Practice: Responding to the Risks Posed by Climate Change: Cities Have No Choice But to Adapt." *Town Planning Review* 81 (3): 217–35.

Susskind, L., and Field, P. 1996. *Dealing with an Angry Public: The Mutual Gains Approach to Resolving Disputes*. New York: Free Press.

Susskind, L., Rumore, D., Hulet, C., and Field, P. 2015. *Managing Climate Risks in Coastal Communities: Strategies for Engagement, Readiness and Adaptation*. London: Anthem Press.

Tang, Z., Brody, S. D., Quinn, C., Chang, L., and Wei, T. 2010. "Moving from Agenda to Action: Evaluating Local Climate Change Action Plans." *Journal of Environmental Planning and Management* 53 (1): 41–62.

Thompson, J. P., III. 2005. *Double Trouble: Black Mayors, Black Communities, and the Call for a Deep Democracy*. New York: Oxford University Press.

Tol, R. S. 2003. "Is the Uncertainty about Climate Change Too Large for Expected Cost-Benefit Analysis?" *Climatic Change* 56 (3): 265–89.

Tsai, F.-H., Kuang-Chao, Y., and Hsiao, H.-S. 2012. "Exploring the Factors Influencing Learning Effectiveness in Digital Game-Based Learning." *Journal of Educational Technology & Society* 15 (3): 240–50.

Turner, L. R., Alderman, K., Connell, D., and Tong, S. 2013. "Motivators and Barriers to Incorporating Climate Change-Related Health Risks in Environmental Health Impact Assessment." *International Journal of Environmental Research and Public Health* 10 (3): 1139–51. https://doi.org/10.3390/ijerph10031139.

Tversky, A., and Kahneman, D. 1981. "The Framing of Decisions and the Psychology of Choice." *Science* 211 (4481): 453–58.

U.S. Census Bureau. 2015. *2011–2015 American Community Survey 5-year Estimates*. Retrieved from https://factfinder.census.gov/faces/tableservices/jsf/pages/productview.xhtml?src=CF.

Uejio, C. K., Wilhelmi, O. V., Golden, J. S., Mills, D. M., Gulino, S. P., and Samenow, J. P. 2011. "Intra-Urban Societal Vulnerability to Extreme Heat: The Role of Heat Exposure and the Built Environment, Socioeconomics, and Neighborhood Stability." *Health & Place* 17 (2): 498–507.

Union of Concerned Scientists. 2019. "Killer Heat in the United States: Climate Choices and the Future of Dangerously Hot Days." July 2. Retrieved from https://www.ucsusa.org/global-warming/global-warming-impacts/killer-heat-in-united-states.

United States Global Change Research Program (USGCRP). 2014. *Climate Change Impacts in the United States: The Third National Climate Assessment*. U.S. Global Change Research Program, Washington, DC. Retrieved from https://nca2014.globalchange.gov.

———. 2016. "The Impacts of Climate Change on Human Health in the United States: A Scientific Assessment." Global Change Research Program, Washington, DC. Retrieved from https://health2016.globalchange.gov/.

———. 2018: *Impacts, Risks, and Adaptation in the United States: Fourth National Climate Assessment, Volume II*. U.S. Global Change Research Program, Washington, DC. Retrieved from https://nca2018.globalchange.gov/.

van der Linden, S., Leiserowitz, A., and Maibach, E. W. 2016. "Communicating the Scientific Consensus on Human-Caused Climate Change Is an Effective and Depolarizing Public Engagement Strategy: Experimental Evidence from a Large National Replication Study." *SSRN*. Retrieved from https://papers.ssrn.com/sol3/papers.cfm?abstract_id=2733956.

Vargo, J., Stone, B., Habeeb, D., Liu, P., and Russell, A. 2016. "The Social and Spatial Distribution of Temperature-Related Health Impacts from Urban Heat Island Reduction Policies." *Environmental Science & Policy* 66: 366–74.

Vogel, J. J., Vogel, D. S., Cannon-Bowers, J., Bowers, C. A., Muse, K., and Wright, M. 2006. "Computer Gaming and Interactive Simulations for Learning: A Meta-Analysis." *Journal of Educational Computing Research* 34 (3): 229–43.

Vogel, J., Carney, K. M., Smith, J. B., Herrick, Stults, M., O'Grady, M., Juliana, A. S., Hosterman, H., and Giangola, L. 2016. *Climate Adaptation: The State of Practice in U.S. Communities*. Kresge Foundation, Detroit. Retrieved from https://kresge.org/sites/default/files/library/climate-adaptation-the-state-of-practice-in-us-communities-full-report.pdf.

Watts, N., Adger, W. N., Agnolucci, P., Blackstock, J., Byass, P., Cai, W., Chaytor, S., Colbourn, T., Collins, M., Cooper, A., Cox, P. M., Depledge, J., Drummond, P., Ekins, P., Galaz, V., Grace, D., Graham, H., Grubb, M., Haines, A., Hamilton, I., Hunter, A., Jiang, X., Li, M., Kelman, I., Liang, L., Lott, M., Lowe, R., Luo, Y., Mace, G., Maslin, M., Nilsson, M., Oreszczyn, T., Pye, S., Quinn, T., Svensdotter, M., Venevsky, S., Warner, K., Xu, B., Yang, J., Yin, Y., Yu, C., Zhang, Q., Gong, P., Montgomery, H., and Costello, A. 2015. "Health and Climate Change: Policy Responses to Protect Public Health." *The Lancet* 386 (10006): 1861–914. https://doi.org/10.1016/S0140-6736(15)60854-6.

Weber, E. U. 2006. "Experience-Based and Description-Based Perceptions of Long-Term Risk: Why Global Warming Does Not Scare Us (Yet)." *Climatic Change* 77 (1–2): 103–20.

———. 2010. "What Shapes Perceptions of Climate Change?" *Wiley Interdisciplinary Reviews: Climate Change* 1 (3): 332–42.

Weber, E. U., and Stern, P. C. 2011. "Public Understanding of Climate Change in the United States." *American Psychologist* 66 (4): 315–28.

White, S. C. 1996. "Depoliticising Development: The Uses and Abuses of Participation." *Development in Practice* 6 (1): 6–15.

Whitmarsh, L., O'Neill, S., and Lorenzoni, I. 2013. "Public Engagement with Climate Change: What Do We Know and Where Do We Go from Here?" *International Journal of Media & Cultural Politics* 9 (1): 7–25.

World Health Organization (WHO). 2013. *Protecting Health from Climate Change: Vulnerability and Adaptation Assessment*. World Health Organization. Retrieved from https://apps.who.int/iris/bitstream/handle/10665/104200/9789241564687_eng.pdf?sequence=1&isAllowed=y.

Williams, H. T., McMurray, J. R., Kurz, T., and Lambert, F. H. 2015. "Network Analysis Reveals Open Forums and Echo Chambers in Social Media Discussions of Climate Change." *Global Environmental Change* 32: 126–38.

Williams, M. K., Green, A., and Kim, E. 2017. "Municipal Leadership of Climate Adaptation Negotiations: Effective Tools and Strategies in Houston and Fort Lauderdale." *Negotiation Journal* 33 (1): 5–23.

Winn, J. 2003. "Avoiding Death by PowerPoint." *Journal of Professional Issues in Engineering Education and Practice* 129 (3): 115–18.

Wondolleck, J. M., and Yaffee, S. L. 2000. *Making Collaboration Work: Lessons from Innovation in Natural Resource Management*. Washington, DC: Island Press.

Woodruff, S. C., and M. Stults. 2016. "Numerous Strategies but Limited Implementation Guidance in US Local Adaptation Plans." *Nature Climate Change* 6 (8): 796–802.

Wouters, P., Van der Spek, E. D., and van Oostendorp, H. 2009. "Current Practices in Serious Game Research: A Review from a Learning Outcomes Perspective." In *Games-Based Learning*

Advancements for Multisensory Human Computer Interfaces: Techniques and Effective Practices, edited by T. Connolly, M. Stansfield, and L. Boyle, 232–55. Hershey, PA: Information Science Reference.

Wright, G., and Goodwin, P. 2002. "Eliminating a Framing Bias by Using Simple Instructions to 'Think Harder' and Respondents with Managerial Experience: Comment on 'Breaking the Frame'." *Strategic Management Journal* 23 (11): 1059–67.

Wu, J. S., and Lee, J. J. 2015. "Climate Change Games as Tools for Education and Engagement." *Nature Climate Change* 5 (5): 413–18.

INDEX

adaptation literacy 41
air quality/pollution, health impacts of 8t2.1, 63, 92t8.2
American Heart Association 7
American Medical Association 7
Arnstein, S. R. 22
Aslin, H. 20
attitude 66n2
  changes in 79–80
  definition of 45n2

Bebbington, A. 22, 83
Beierle, T. C. 23
Biggs, S. D. 22
BRACE framework (Building Resilience Against Climate Effects) 10, 11t2.2, 12
Brown, V. 20
Building Resilience Against Climate Effects. *See* BRACE framework (Building Resilience Against Climate Effects)

California Health and Equity, framework for health-climate linkage 11, 12, 12t2.3
Centers for Disease Control and Prevention (CDC)
  BRACE framework (Building Resilience Against Climate Effects) 10, 11t2.2, 12
Ceyhan, P. 62
Clim'way 62
climate adaptation 1–2, 95
  definition of 17
  framing of 18–19
  planning and policymaking 2
    and health inequalities 10
    in municipality 2
    operational process of 17
  planning
    definition of 17
    recommendations for 9
  shortcomings of efforts 9, 11
  social barriers to 41
climate change
  framing as public issue 35
climate change, framing as public health issue 31–32, 37–38, 48, 87–88
  and public opinion 29–30
  effects of 34t4.1
  efficacy of 35–37, 38
  emphasis framing effect 30
  equivalency framing effect 29
  frames 30–31
  research process 32–33
  research results 33–37
  serious games. *See* digital games; role-play simulations
Climate Kids 62
climate policy agenda 7
co-benefits 49
collaborative capacity 41
collective climate adaptation, at local level 23
  barriers 18
  framing of 18–19
  planned adaptation 17
  and policymaking 19–23
collective problem-solving 96
community-based action 21
consultation fatigue 21
conventional public meetings 61

digital games 76, 88–89, 95
  and real-life situations 70
  and climate change 62
  climate knowledge, affect, and behavior by political viewpoint, shift in 68

digital games (*cont.*)
climate knowledge, affect, and behavior, shift in 67–68
definition of 61
effects of 68t6.1, 69t6.2, 70f6.5, 70t6.3, 81t7.2, 81t7.3, 82t7.4
efficacy of 61
on health impacts of climate change; health impacts 62–64
inside experience 74–75
learning types in 69, 70t6.3
local climate adaptation policy priorities, shifts in 70
as policy recommendation 93–94
prioritization of local climate risks in climate adaptation 71t6.4
proactive learning 72f6.6, 72f6.7, 73–74
promise of 61
research process 64–66
research results 66–72
versus role-play simulations 84t7.7, 89, 90t8.1
advantages and disadvantages 83–85
attitudinal changes 79–80
demographics 79, 80t7.1
learning types 82, 83t7.5, 83t7.6
Dilling, L. 30

education 73
emergency preparedness 47, 54, 55, 74, 76
emphasis framing 30, 32
Entman, R. M. 29
equivalency framing effect 29
extreme heat, health impacts of 8t2.1, 92t8.2
extreme weather events, health impacts of 8t2.1, 92t8.2

face-to-face role-play simulation. *See* role-play simulations
Farrington, J. 22, 83
fecal contamination 1
Few, R. 22
Fischhoff, B. 30
Fourth National Climate Assessment 17
Frames and Games project 2, 12, 23, 37, 43, 87, 88
Frewer, L. J. 22

Graesser, A. 61
greenhouse gas emissions 1

health impacts of climate change 2, 7, 11–12, 92t8.2, 95. *See also* climate change, framing as public health issues
BRACE framework (Building Resilience Against Climate Effects) 10, 11t2.2, 12
California Health and Equity's framework 11, 12, 12t2.3
lack of knowledge on 9
planning for and management of 7–11
health organizations, climate policy agenda 7
HeatReady City program 9
Hong, H. 62
Hot Weather-Health Watch/Warning System 9
Hovland, C. I. 45n2, 66n2
Hughes, S. 18, 20, 23
human vulnerabilities 2, 19, 21
Hurricane Harvey 1

information provision 73
Intergovernmental Panel on Climate Change (IPCC) 17, 19, 20
Special Report on Emissions Scenarios 20

Jamieson, S. 34n9, 67n7
Jordan-Cooley, W. 62

Kahneman, D. 29
Kim, T. 62

Lee, J. J. 62
local climate adaptation plans and policies 90
long-term community needs, planning for 48
Los Angeles, climate adaptation in 9
low-income and vulnerable populations 48

Magerko, B. 62
Mandanaro, L. A. 23
Miami-Dade County 18
Moser, S. C. 23, 30
Myers, T. A. 32

NASA, *Climate Kids* 62
Nelson, T. E. 30
New England Climate Adaptation Project (NECAP) 43
New York City 18
Nisbet, M. C. 30, 31

online engagement 18, 49

parametric analyses 67n7
Philadelphia, climate adaptation in 9
Phoenix, climate adaptation in 9
Pike, C. 23
planned adaptation to climate change 17
playing games, learning outcomes 42
policy recommendations
  digital games deployment 93–94
  framing climate change as a health issue 90–91
  order of serious games deployment 95
  role-play simulations deployment 91–93
political orientation 51
Pretty, J. N. 22
priming 30
proactive learning 72f6.6, 72f6.7
problem identification 90
public engagement 19–23, 41, 96
  online. *See* digital games
  outcome evaluation 23
  role-play simulations' impact on 48
public health frame. *See* climate change, framing as public health issue
public information 49
public opinion 20
  on climate change 48
  manipulation of 31
  and policy issue framing 29–30

renewable energy 74
Renn, O. 22
role-play simulations 41, 57, 88, 95
  climate risks management 55t5.5
  definition of 42
  design of 43n1
  versus digital games 84t7.7, 89, 90t8.1
    advantages and disadvantages 83–85
    attitudinal changes 79–80
    demographics 79, 80t7.1
    effects of 82t7.4
    learning types 82, 83t7.5, 83t7.6
  effects of 50, 51t5.1, 52t5.2, 53t5.3, 54f5.2, 81t7.2, 81t7.3
  on health impacts of climate change 43–44
    debriefing sessions 48–49
    group decision-making success factors 49–50
    research process 44–45
    research results 47–57
  inside experience 56–57
  learning types in 53
  local climate adaptation policy priorities, shifts in 54
  local climate risks
    management policy, shifts in 54–55
    prioritization 55t5.4
  in planning and policymaking context 42
  as policy recommendation 91–93
  shifts in knowledge, affect, and behavior by political viewpoint 51–53
  as tools for education and public engagement 42
Rosenberg, M. J. 45n2, 66n2
Rowe, G. 22
Rumore, D. 41

Sailor, D. 7
Schenk, T. 41
Scheufele, D. A. 30, 31
serious games 41–43. *See also* digital games; role-play simulations
short-term public health impacts of climate change 48
Sniderman, P. M. 30
social barriers to climate adaptation 41
Stern, P.C. 30
Stone, B. 7
Sung, W. 62
Susskind, L. 17, 20, 22, 41

temperature increase, health impacts of 8t2.1, 92t8.2
temperatures 7
Theriault, S. M. 30
Third National Climate Assessment 17
tripartite model of attitude structure 66n2
Tversky, A. 29

Urban Cooling Collaborative 9
urban cooling strategies 9, 10
U.S. Environmental Protection Agency (EPA) 63

Weber, E. U. 30
White, S. C. 22
wildfires, health impacts of 8t2.1, 92t8.2
World Health Organization (WHO) 7
Wu, J. S. 62

www.ingramcontent.com/pod-product-compliance
Ingram Content Group UK Ltd.
Pitfield, Milton Keynes, MK11 3LW, UK
UKHW041832100926
8556IPUK00002B/125

9 781785 273247